The Strangest Things

A Book About Extraordinary Manifestations of Nature

Thomas R. Henry

Alpha Editions

This edition published in 2024

ISBN : 9789362990655

Design and Setting By
Alpha Editions
www.alphaedis.com
Email - info@alphaedis.com

As per information held with us this book is in Public Domain.
This book is a reproduction of an important historical work. Alpha Editions uses the best technology to reproduce historical work in the same manner it was first published to preserve its original nature. Any marks or number seen are left intentionally to preserve its true form.

INTRODUCTION

The challenges of Nature's paradoxes have been sharp spurs to man's search for knowledge since the start of science.

Fortunately the number of these paradoxes is infinite, and so the quests are endless. Man never will know a wonderless world. In the phenomena of life especially we have come only to the zone of morning twilight. The bright day of understanding is ahead. As its hours pass we can expect a constant succession of new paradoxes, new spurs to further advances.

Man would be in a sad situation were it otherwise. For the bright light of noon and afternoon inevitably precedes sunset and darkness and sleep.

This book is a compendium of some of Nature's curiosities and contradictions in the field of life and as such it well may awaken that wonder which, as somebody has said, is the beginning of knowledge.

The author is one of the world's best-known and most respected science writers. This book is a personal and unique distillation of the wisdom he has developed in a lifetime of dealing with man's effort to resolve the paradoxes of nature.

Leonard Carmichael

Secretary of the
Smithsonian Institution

PREFACE

Life has invaded nearly every crack and crevasse of the world during the billion years since it left its first traces on this planet. It has adjusted itself to all extremes of living, from nearly airless mountaintops five miles high to lightless floors of oceans five miles deep. It has found abodes in boiling hot springs and in the everlasting ice of Antarctic peaks. It very likely has invaded the cold, red deserts of Mars. Everywhere it has succeeded in altering the garments it wears to meet the stresses it has experienced.

It has achieved semi-infinite variety. There are approximately a quarter million species of plants now known in the world. Most abundant and varied life is that of the insects who may be on their way to displace man and his fellow mammals as lords of the earth. A rough estimate of the number of species identified up to now is 800,000. Several thousand hitherto unknown are described each year. Of mammals, including man, there may be as many as 14,000 distinct species and geographic races extant. About 8,500 species of birds are catalogued. Sub-species and geographic races increase this number to about 30,000. Known fishes number 40,000 species and sub-species.

Still, naturalists say, there are great mansions of life almost unknown to man. The collections of the Smithsonian Institution in Washington grow at the rate of about a million specimens a year, always including forms hitherto uncatalogued. Much of the material in the following pages is based on Smithsonian information, although other sources and personal observations have been liberally drawn upon.

The Smithsonian specimens, as well as those in other museums and collections throughout the world, are types. Once they were individuals with passions, fears, hungers, perhaps some dim wonderings and questionings. The type is the eternal reality. The individual is the brief-lived example of this reality, the flame of a candle fluttering in a windy moment.

I have brought together in these pages notes about the most extraordinary manifestations of nature that have come to my attention in the course of thirty years as a science reporter. Each example is, of course, based upon a distinctly individual expression of nature, but all are very much interrelated in this truly amazing world of ours.

THOMAS R. HENRY

Washington, D. C.

THE INVISIBLE UNDERGROUND JUNGLE

There may be as many as twenty-five million invisible plants and animals in a gram of soil about the size of a grain of sand. It would take a thousand such grains to make a marble.

The population of this microscopic jungle is composed chiefly of single-celled organisms—bacteria, molds, yeasts and protozoa. Total numbers vary enormously—from time to time and place to place—chiefly because of variations in the food supply. Although thousands of species have been identified, the greater part of soil life still remains unknown.

This jungle is a place of the hunter and the hunted—of an incessant and merciless struggle for survival. Invisible plants eat invisible animals and invisible animals eat invisible plants. Plants devour other plants and animals devour other animals.

Giants of this nether world—largely invisible, although the average size is more than a thousand times that of the bacteria—are thread-like white worms from a hundredth to a fifth of an inch long. Relatively they are not very plentiful—less than six million to a cubic foot of soil in most places. In both size and numbers in the earth population, they are like elephants compared to mice. Still they probably are numerically the most abundant of all animals which consist of more than a single cell. In the entire animal kingdom only the protozoa outnumber them.

These creatures are the nematodes, or eel worms. About ten thousand kinds have been described; there are probably as many more unknown to zoologists. Less than a hundred of these varieties cost American farmers and gardeners more than half a billion dollars a year. The rest of those species living in the soil are, so far as known, harmless or even slightly beneficial. Seas and fresh waters are full of other kinds. Still others, some very much larger than the soil organisms, are among the most dangerous parasites of animals and men. The little soil worms, in the opinion of Dr. Geoffrey LaPage of Cambridge University, "must be considered one of the major menaces of our civilization."

Although always invisible, the activities of these countless billions of organisms underfoot can be measured in various ways. For example, carbon dioxide is constantly escaping from the surface of the ground. This comes from the breathing of the unseen animals and plants. Measurement of the gas outflow gives a rough estimate of how many are present. It shows that the numbers vary greatly from hour to hour.

The soil organisms are relatively immune to heat and cold, flood and drought. Even when a grain of soil has been made absolutely dry in the laboratory and then crushed to a very fine powder, they still remain. If it is placed in a sterile container filled with some fermentable material, a seething mass of microörganisms will appear in a few hours.

Some day this vast, unseen mass of life may be harnessed to the service of man. Only beginnings have been made to achieve this end. Some of the microscopic life forms are definitely helpful to plant life, while others undoubtedly are destructive. One service, without which plant life would be unable to continue very long, is the fixation in the soil of nitrogen from the air. One group of bacteria, the azotobacteria, do this in the laboratory and long have been supposed to be the effective agents in nature. But actual examination of soil samples, say Department of Agriculture specialists, fail to show more than a few thousands of these organisms per gram of soil anywhere, and sometimes none at all can be found in places where it is known that nitrogen fixation is in progress. Some still unknown form of microscopic life must be doing part of the work.

Another unknown organism is an agent partly responsible for breaking down the cellulose of dead plants in the soil. The mold, Aspergillus fumigatus, world-wide in its distribution, does this in the laboratory. Nowhere, however, is it found in nature in sufficient numbers to accomplish the titanic job attributed to it.

The great, invisible jungle, of course, must eat to live. Some organisms demand fresh food and are responsible for root rot in plants. The majority, however, find their sustenance in the enormous mass of dead and dying roots of annual vegetation. Decomposition of annuals is an explosive process involving the development of countless billions of bacteria.

THE SELF-PERPETUATING SPONGE

Close to primaeval chaos is the sponge—lowliest of animals. It is an animal without a brain, nervous system, heart, lungs, stomach, muscles or blood. But it has an *I Am*.

The sponge is in essence an anarchical horde of numberless cells. When the conglomeration is split up as can be done by a technique of squeezing through fine-meshed silk gauze, the cells continue to live as individuals. They crawl about. They take nourishment. But when a few thousands of them are thrown together into a tank of sea water they will conglomerate again, apparently into the same sponge that existed before the disintegration. If sponge animals of two different species are mixed in the tank they will combine into two sponges, duplications of the conglomerations from which they came. If cells of two sponges of the same species are mixed it may be that they will recombine into the two original individuals—but this experiment never has been tried and would be quite difficult to interpret.

The sponge is the simplest, most primitive of metazoa, or many-celled animals. It acts as an individual, although there is apparently no central government, like a brain, controlling the behavior of the millions of individuals constituting the conglomeration. It ranges in size from organisms a fraction of an inch long, by far the most numerous, to masses several feet in diameter. Various species present about all the colors of the rainbow. There are red, scarlet, green, yellow, blue and violet sponges, especially in shallow, tropical waters. Abysmal species tend to be a drab brown.

The living sponge when taken from the water is a slimy, rather repulsive mass which has the general appearance of a piece of raw beef liver perforated with holes and canals. The commercial sponge is merely the skeleton, the supporting framework of the gelatin-like tissues, which is composed of a substance similar in chemical and physical properties to silk, horn and the chitin which forms the shells of insects and crabs. This material is distributed in a fibrous network the pattern of which varies for each species.

The sponge has the most remarkable powers of regeneration of lost parts known in nature. It can regrow its entire body from a small fragment of itself. Thus if a sponge were cut into fine parts and each fragment cemented to a bit of rock each would grow into a complete, normal animal. Also if a sponge is cut or torn away from the sea bottom in such a way that some fragment remains attached this fragment will continue growing.

LIVING "STARS" IN CAVES

There is a cathedral-like grotto under the earth whose roof is lit eternally by living stars. It is an enormous labyrinthine chamber cut by a slow-flowing river in the base of a limestone mountain.

Its dome is like the dome of the heavens on a frosty October night. There shine the Big Dipper, the Southern Cross and the Belt of Orion. The Clouds of Magellan are on the southern horizon. There are millions of pale stars grouped in all sorts of astrological configurations. Some are isolated in space. Some are packed in dense galaxies. There are black voids between them, like the curtain of star dust that hides the center of the universe. They are only a few feet overhead. One can reach up and pluck these stars, one by one, out of the sky. Unlike the heavenly bodies, they do not twinkle. They shine steadily in complete motionlessness. Pale and weird, they illumine a realm of eternal night. It is a domain of absolute silence. Around the walls the strange starlight falls on carved figures of winged angels, of human faces laughing and human faces contorted in agony. Each star is a predacious living animal, a flesh-hungry hunter and killer. From it is suspended four or five foot-long strings of shining pearls, so delicate that they shimmer at a human breath.

This star-lit cave near the little city of Te Awaamutu is New Zealand's greatest curiosity and certainly one of the weirdest and most intriguing spots on earth. The grotto constitutes about a third of the Waitome caverns in the center of Maoriland in the North Island, otherwise rather featureless, water-chiseled rooms in the depths of a mountain with the customary stalagmite and stalactite formations.

The stars are luminous, slimy, dirty-grey worms. They are rarely found anywhere else, and never in very great numbers. This is the one spot on earth ideally adapted to their unbelievably queer life cycle. The worm is the larva of a dainty, dark-winged fly about twice as large as a mosquito, which looks like a miniature daddy longlegs. It has no common name. Scientifically it is classified as Boletophela luminosa, a member of the sub-order of arachenocampa. It falls somewhere between true insects and spiders. There is no relationship between it and any other luminous insect—glow-worm or firefly—anywhere.

The light is a lure for prey to satisfy a voracious appetite. The lovely strings of pearls are modifications of the spider's web. Nature has provided few other creatures with so intricate and ingenious a food-gathering mechanism as that which enables this *none* to survive in its strange environment Here evolution has schemed in an unique way to ensure the preservation of a species which apparently serves no purpose in the economy of nature except to procreate a beauty spot

The floor of the glow worm grotto is a subterranean branch of a river. The water is warm and almost absolutely motionless, for no breezes penetrate that far under the mountain. Thus it is an almost ideal spot for all sorts of insects to lay their eggs. There is a high probability that the great majority of them will hatch. As the young rise from the water they are attracted by the star-filled heavens overhead. They fly toward them as moths to a lamp. The same is true of many of the small adult insects, some of which are essentially microscopic. Once such an insect is caught on one of the threads it is lost beyond all hope. There it sticks, struggle as it may. The vibrations caused by its struggles attract the attention of the glow worm which quickly winds up the hanging thread. If it is not hungry at the moment it has been observed to play with its victim, drawing in and then letting out the line after the manner of a fisherman. Finally the prey is drawn into the silken sheath and entirely devoured, chitinous shell and all. It is not merely sucked, as is the fashion of the spider or the fly.

The "lamps" apparently are under an extremely delicate nervous control. The strings of pearls suspended loosely in the air must be extraordinarily sensitive to sound waves. The instant they pick up any sound unusual for the cavern the lights automatically go out. Stranger still is the fact that the darkening of all the stars is nearly simultaneous. This, of course, is a safety measure. Any disturbance of the cave routine means danger for the transparent caterpillars. In order to see the star-lit heavens effect the row boat in which one enters the glow worm grotto must be handled by skilled oarsmen so that there is no sound of splashing water. Visitors are warned not even to whisper, lest some string be disturbed and instantaneously transmit the warning to all the others.

PARENTHOOD AMONG PENGUINS

One of nature's miracles is the egg-laying and incubating of the emperor penguin in the darkness of the Antarctic night at temperatures of from 50 to 80 degrees below zero.

Dr. Edward Wilson, surgeon of Sir Robert Falcon Scott's 1901 south polar expedition, found the first emperor rookery and was able to observe it for several days. His account became one of the classics of science. The big birds hatched their eggs, he found, standing on one foot on the ice and holding them against the breast feathers with the other foot. The task evidently was shared by both males and females. The male would take the egg from the female while she trekked to open water to feed on fish. After a few days, Wilson supposed, she would return while the male went after fish.

In 1956 Dr. Bernard Stonehouse of the Falkland Island Dependencies Administration found another emperor rookery and maintained observations for about ten weeks. The behavior observed was even more of a miracle than Dr. Wilson supposed.

After laying their eggs on the ice, Stonehouse noticed, the females leave immediately for open water and remain there for sixty days, the full period of incubation. Presumably they feed constantly during this period. The males take over entirely at the rookery. For two months the husband remains standing on one foot and holding an egg against his breast with the other—presumably shifting his feet now and then. Through the entire hatching period he eats nothing. When the eggs are about to hatch the mothers return from the sea, tidy up the nursery, and get ready to take over rearing the chicks. Then the males, who have exhausted their reserve of fat, stagger feebly in their own mass migration to open water to rebuild their reserves on fish. By the time of the Antarctic sunrise in October the chicks are about ready to fend for themselves.

Standing from three to four feet high and looking and acting deceptively like a human being, the emperor penguin undoubtedly is one of the most remarkable birds in existence. It presumably is confined to the Ross Sea side of the Antarctic continent. The bird—actually it is about two-thirds feathers—remains an evolutionary enigma. Theories have been advanced that it is the last surviving member of the fauna of the Antarctic continent about fifty million years ago when the shorelines were free of ice. It certainly is off any known road of evolution.

THE STRATEGY OF WARRIOR ANTS

Total war is the way of life for army ants. The picturesque, devastating drives of their vast hordes have nothing whatever to do with exhaustion of food or anything of the sort. The wars come in fixed cycles, regardless of supplies.

There are two species of these ants on Barro Colorado Island in the Panama Canal Zone. Each species has approximately 50 colonies and each colony consists of from a few hundred thousands to more than a million individuals. At the head of each colony is a single queen who lays all the eggs.

There is a new lot of larvae every 33 days—all workers or incompletely developed females. Development is restricted by the amount of food available. Since each brood consists of about 60,000 individuals, a colony theoretically might reach titanic proportions. However, it does little more than maintain its population. The death rate of soldier ants, in constant combat, is very heavy.

Once each year, at the start of the dry season in the tropics, a colony queen produces a sexual brood of about 3,000 males and six queens. The rest of the 60,000 eggs laid at this time are incapable of hatching and are fed to the new-born sexed individuals. They apparently have some of the nutritious properties of the royal jelly fed to queen bees.

This sexual brood is produced in what has been called a statory period in which the army maintains a fixed bivouac for about three weeks. During this time the new queens develop and around at least one of them a new group of workers, about half the whole, tends to congregate. A strange antagonism seems to develop between the old and new groups. Eventually the colony divides in two and each half starts moving in opposite directions. The other new queens are lost in the shuffle.

Most of the newly developed males are 'excess baggage.' During the winged, or mating, stage they fly into the forest where the great majority of them are eaten by birds. When the surviving ants light on a tree, on the ground or on some other object, the wings drop off. Then they apparently wander about aimlessly until they come to an army ant trail which they recognize by the odor and follow it until they come to the colony which has made it. If this happens to be a colony of their own relatives, they probably are killed by the workers. If it happens to be an entirely foreign colony, they may be accepted. This apparently is one of nature's mechanisms for intruding new genes into a strain.

The raiding activities of a colony are carried out during the day from a central headquarters. During the daytime raiding individuals return to the colony from their forays and by dusk all have returned. At night the bivouac is changed, the whole colony moving forward along one of the trails blazed by the raiders. A new headquarters is thus established. A colony moves from six to seven hours before striking a new bivouac. Not infrequently, if no promising site is found, it moves from dusk to dawn.

This would seem like constant activity, too strenuous even for the constitution of an army ant. Actually the individual workers probably get plenty of rest. Each colony is divided into two units—the raiders and those that constitute the structural unit. The walls of the "headquarters" are made up of the bodies of the latter. These "living brick" do nothing throughout the day. They may be asleep. When the raiders return at dusk the structural unit breaks up and the members lead the migration to a new bivouac. The erstwhile raiders follow leisurely in the rear and in turn become the structural unit when a stopping place is selected.

When to rest? When to raid? There apparently is an irresistible war rhythm, like the rhythm of the tides, in the basic constitution of these ants. Some have postulated the same sort of thing, on a lesser scale, in man who goes to war every so often but camouflages the war tide with economic or political explanations.

These ants are remarkable not only as warriors but as architects. They build complex, air-conditioned, hanging houses out of thousands of their own suspended bodies. Within these structures the queen is sheltered, eggs laid, young hatched and reared. Much of the time the "houses" are constructed anew each night.

This home-building behavior is unique in nature, as Dr. T. C. Schneirla of the American Museum of Natural History has pointed out:

"Without any active excavating and without any manipulating of fallen materials, colonies of these species form a domicile with their own bodies. A typical bivouac is a cylindrical mass hanging from the underside of some projecting surface to the ground. In addition to the sides or under-surface of logs, other typical places are the spaces between gut tressed tree roots, masses of brush, undercut banks of stream beds, or the overhanging edge of a rock.

"The characteristic ability to cluster their bodies, as well as the manner of clustering, depends first of all upon an anatomical characteristic—the opposed, recurved hooks on the terminal tarsal segments of the workers' legs. The first ants to settle in a new place catch into a rough or soft surface by means of the tarsal hooks, or rather are pushed into this anchored position as newcomers run upon them as they stand and stretch them out in a hanging position. In fact, the hooks are really anchored by the added weight of others that have crawled down over the body of the first ant, fixing it in place and soon immobilizing it.

"In the nomadic phase a new bivouac is formed at the end of each day of raiding. In the advanced and most complicated stages of raiding in the afternoon, caches of booty tend to be formed at each busy junction of raiding trails, increasing in size as more and more ants are knocked around and forced out of traffic. As darkness comes and raiding ceases such clusters grow. Several hanging clusters start from elevated ceilings. As each new cluster begins, the initial slender hanging threads may become ropes which extend to the ground. As the ropes continue to grow they are joined together into a single columnar mass.

"At first this mass is small in diameter, but as more and more ants pour into it the wall spreads outwards from the center and so a symmetrical cylinder results."

In the tropical environment of the army ants some sort of air conditioning is necessary for comfortable living—perhaps, with this particular species, for any living at all. It has been well developed during the more than 50 million years the insects have been on earth. Says Dr. Schneirla:

"The interior of the bivouac, where the brood is sheltered and the single colony queen rests, offers an impressively stable environment to these more susceptible members of the community as well as a central resting place for the worker population. The hanging cluster traps a cubic area for atmosphere which does not reach the extremes of temperature and dryness attained by the general forest environment, but in general is somewhat warmer and more humid at night and somewhat cooler and dryer during the day.

"This result is achieved mainly as a result of worker behavior. Workers cluster more closely together at night in reaction to the lower temperature of the forest at the time. The bivouac walls become tighter and thus better conserve heat produced internally by the brood.

"Conversely, after dawn, when increasing light excites growing numbers of ants to leave the bivouac, as the raid grows, this wall thins out, usually develops small apertures, and is undercut at the bottom.

The effect is to increase internal air circulation as well as to cool the atmosphere of the interior through evaporation, so that the internal temperature of the bivouac does not rise to the height reached at midday in the environs.

"The incubation properties of the bivouac represent an important factor in echelon life, for with less regular atmospheric conditions in the nest the stages of brood development could not have their typical regularity in timing."

UGANDA'S MINIATURE DINOSAUR

A grotesque creature abundant in the Kishasha Valley of Uganda is the three-horned chameleon. It grows to a length exceeding twelve inches and the males look like miniature versions of the ancient dinosaur monster, triceratops. Three curious horns, an inch to an inch-and-a-half in length, protrude from the nose and between the eyes of males.

These are extremely pugnacious animals; they use their horns in fights to the finish. At times the contests develop into prolonged pushing matches with the horns interlocked, but a really vigorous fighter can dispose of an adversary in a few minutes. African natives are terrified of these demoniacal-looking little animals.

THE STRANGE WAYS OF SPIDERS

"With other classes of animals, and even with plants, man feels a certain kinship—but spiders are not of his world. Their strange habits, ethics and psychology seem to belong to some other planet where conditions are more monstrous, more active, more insane, more atrocious, more infernal than on our own. Frightfulness and ruthlessness appear a part of their nature and we stand appalled when it dawns upon us that they are far better armed and equipped for their life work than we for ours."

Thus writes Dr. W. E. Stafford, U. S. Department of Agriculture naturalist. There probably is quite general agreement with his sentiments.

One chills at the picture of some other planet where spiders and their kin who have evolved minds equal to that of humans are the dominant animals.

Once gigantic spider-like creatures ruled this world. They were as big as lions or gorillas. Their realm was the earth of the Silurian geological era of 350,000,000 years ago—a time of warm, quiet seas which, especially in the northern hemisphere, covered large areas that now are dry land. These creatures were the euripterids, or sea scorpions, whose nearest extant relatives are the horseshoe crabs with sky-blue blood that are common along the Atlantic coast of the United States, and the venom-fanged land scorpions. They exceeded in size all living invertebrate animals.

Many were five to six feet long; one was nine feet long. Presumably they were free-swimming, predacious creatures with massive, crushing jaws. Their chief prey, it is believed, were the much smaller, crab-like trilobites with whom they shared a common ancestry. These were shelled animals the imprints of whose hard shells in mud (which later became rock) are among the most ancient records of animal life on this planet. The trilobites were creatures who crawled on shallow sea bottom. Their only defense was to roll themselves in balls. They appear to have been the dominant form of life for at least 100,000,000 years. They continued a precarious existence after the evolution of the great pseudo-spiders, but were well on their way to extinction. The massive jaws of the euripterids could crush their thin shells with ease. The dominance of these new masters of the sea would be challenged only by the gigantic mollusks, but for many millenia they appear to have held their own against these frightful monsters.

Their decline had started by the end of the Silurian period and they were extinct in another hundred million years. The reason for their decline is unknown, but perhaps it was related to some decided change in temperature and distribution of the waters. Remarkably well preserved remains of the monsters have been found imbedded in limestone on Oesel Island, in the Baltic. During the Silurian era life was just starting to emigrate from the oceans and establish a precarious foothold on land. Among the earliest land fossils are those of small scorpions, distantly related to the erstwhile master race. The euripterids themselves, however, never tried to leave the sea.

WORMS WITH A THOUSAND EYES

There are worms with a thousand eyes. They are, for the most part, animals of the dank, dark floors of tropical rain forests.

They are narrow, brilliantly colored ribbons of slimy skin which glide at a speed of about six feet an hour over damp moss and leaves in the everlasting twilight. When alarmed they can break up instantly into scores of "blobs of slime" and in a few hours each piece will become a complete new worm. One of them can eat five-sixths of its own body and entirely recover.

These fantastic creatures are the terricola or land planarians—lowliest of worms and one of the lowliest forms of animal life. Only the microscopic protozoa, the slime moulds, the sponges, jellyfish, and corals are more primitive.

They range from fractions of an inch to nearly a foot in length. They are hunters and scavengers. Nearly all are creatures of darkness and dim light—survivors of the haunted dawn of life on earth. They probably are quite close to the ancestral form of all worms. All are free-living animals, although related closely to the degenerate flukes and cestodes, which are internal parasites of man and other animals.

They belong to an enormous clan. There are several hundred known species and perhaps as many more still unknown. These worms are found over most of the world but most abundantly in the damp tropical and sub-tropical rain forests. They are seldom seen in nature although they are fairly well-known in experimental biology classes, for which they are purchased from dealers. Australia has about sixty species. America may have many more, most of which remain undescribed. One would be likely to come upon them only by accident.

Among these land planarians are some of the most fantastic creatures of the animal kingdom. They have been described as "gliding strips of skin." The family includes some of the most brilliantly colored of all living things. They probably represent the earliest traces of eyes and brains in the world.

The "eyes" of the terricola are black dots arranged in two parallel rows along both sides of the back. Some species are two-eyed. Many varieties are eyeless. Hundred-eyed worms are quite common. The black dots are light-sensitive. Presumably they represent the beginning of vision. By means of them the worms can distinguish between light and darkness. They also tell the direction from which light comes. Actually, however, planarians without eyes have the same ability, but they are slower to react. This is demonstrably true for fresh-water forms. For most of the land forms at least exposure to strong sunlight would be fatal.

Each of the eye dots has a nerve connection with the brain. It is quite unlikely, however, that the animals actually see anything, in the sense of discriminating specific objects in their surroundings. In a few species, however, from two to four of these black dots nearest to the brain seem somewhat more complicated than the others. As the faculty of vision evolves among animals these will become actual eyes and all the light-sensitive spots will be discarded. In most planarians, however, the number of eyes increases with the age of the animal.

Nearly all are predatory meat eaters. They are both hunters and scavengers. Some pursue, kill, and eat living animals, such as earthworms and small mollusks, as big as themselves. They apparently are able to locate their victims at some distances by an already evolved sense of smell. One blind Brazilian species is said to pursue earthworms into their burrows several feet underground.

When the victim is overtaken the planarian first enfolds it in its sheetlike, slimy body. Then from its mouth, always on the underside of the body near the middle instead of at the head end, it projects its pharynx, a muscular tube which is part of the digestive system. From this is exuded a substance of some sort which slowly liquifies the flesh. Then the liquid is sucked into the body through the mouth. Digestion then is completed within the digestive tract by special cells which engulf minute particles in the same way as they are engulfed and digested by one-celled animals, the amoeba. The nature of the dissolving material exuded from the pharynx is unknown. It is believed, however, to contain a mixture of enzymes such as those found in the intestinal tracts of higher animals.

Planarians may attack healthy animals and overpower them in spite of their violent struggles against being enfolded in the slimy skin. They are, however, particularly attracted to the sick and injured which they apparently locate by smell. Most of these worms are devourers of dead flesh. A common method of capturing fresh-water forms is to leave a bit of liver or other raw meat exposed in an area they are likely to frequent. Both water and probably land forms will congregate around it. Then the collector is likely to have a difficult job. As the naturalist William Beebe says about one large Venezuelan rain forest species: "To pry one loose and put it in a bottle is like pouring thick, cold molasses mixed with thick glue."

To their activities as scavengers of the forest floor these ancient worms owe their place in the economy of nature. They normally feed several times a week. When kept without food, however, they can stay alive for months. They gradually shrink in size as they digest themselves. The internal organs are reduced little by little as they are absorbed for food. The first to disappear are the reproductive organs. Most planarians have both male and female reproductive systems. Then come the muscles of the body wall. Never however, do the worms eat their own brains or nervous systems, although the brain may be reduced greatly in size. The I Am of the worm can devour its vestments of protoplasm; it cannot eat itself. When food is available again the organs are regenerated and return to normal size.

Instances are recorded where planarians have reduced their length from slightly more than an inch to less than a seventh of an inch in six months.

Closely related to this practice of "eating themselves" is the remarkable ability of the terricolae to break themselves into small fragments each of which will regenerate into a complete worm. This capacity probably has been a major factor in their survival through the aeons since multi-celled life began on earth. What might seem to be their outstanding weakness in the constant struggle for survival—their soft bodies and extremely loose organization—has become their major strength. A planarian can lose at least nine-tenths of its body and still preserve its individual existence. This self-shattering phenomenon constitutes the worm's chief defense in emergencies. It comes into play when any danger threatens. The regenerating ability, especially of fresh-water forms, differs considerably in degree from species to species. Some are unable to regenerate a "brain" out of fragments of the rear part of the body. Complete in every other respect, the remade worms seem incapable of the typical gliding movements of the race. They remain quiet most of the time but can move forward slowly. A tendency to move in circles has been observed. Fragments from the head section, however, quickly become complete animals.

All planarians actually have heads and a "brain," of sorts. The latter consists of two minute bits of nerve tissue just behind the front of the body, oval-shaped and enclosed in a tough capsule. It serves as a center for nerve fibers extending throughout the animal. Here are coordinated the stimuli received from light and heat, and possibly those from odors and sound. When the worm goes forward, it moves its head constantly from side to side. Presumably it is exploring the way ahead for food and danger.

A terrestrial flat worm's progress is described as "gliding," rather than creeping or crawling. The outer surface of the body has many glands from which is exuded a mucus over which it slides. This mucus quickly hardens. From it can be made slender threads by which the worm, like a spider, can lower itself safely from projections. Because of the glue-like quality of the secretion it is able to climb perpendicular surfaces. From the hardened mucous, sometimes mixed with sand, it can make for itself a shell into which it can retire for months at a time.

QUEER FISH, BUT DEFINITELY

There are more than 40,000 kinds of fish in the world. Their habitats range from the profoundest depths of the seas to cold lakes and brooks on mountain timberlines. They show a bewildering diversity in their ways of life.

The smallest of fish is a Philippine goby, less than a third of an inch long and weighing a fraction of an ounce. The largest is the whale shark, found in all warm seas. Some individuals exceed twenty tons.

Some fish burrow in the mud, some swim, some walk, some fly, some breathe air. Some are timid, some bold and bloodthirsty. Some are placid, some easily irritated.

Some are highly venomous. One, found in Australian waters, weighs nearly half a ton and has poison barbs a foot long. Some of the deadliest are among the most beautifully colored.

Freshwater fish can sometimes be cut out of cakes of ice in which they have been frozen for months at a time, and completely revive. Actually the fish themselves are not frozen. The freezing point of their blood is slightly lower than that of water. They were merely "hibernating". This may happen frequently in nature.

Some fish seem well on their way to becoming land animals. They can breathe in air better than in water.

Surgeon fish are so-called because of a sharp spine on the tail which can produce a cut like that made by a surgeon's scalpel.

Parrot fish have beaks like parrots with which they scour algae from the coral reefs for food.

Goat fish have two growths under the mouth which look like the chin whiskers of goats.

Porcupine fish, whose skins are covered with sharp spines and which can fill their sac-like bodies with water or inflate them with air until they form a ball about twice their normal size. When the bodies are puffed up the sharp spines are erected to protect the creatures against their enemies. The inflation is a defense measure which takes place almost automatically when the fish is alarmed.

Trigger fish are creatures with rigid spines which "lock" automatically when the animals are in danger so that they cannot be bent. They can be unlocked, presumably by a nerve reflex, only by the fish themselves or by some scientist who knows the precise spinal process to touch.

Squirrel fish are brilliantly colored little creatures with large deep-brown eyes which look like the eyes of a squirrel.

Scorpion fish have bodies covered with venomous spines whose poison is reputed to be sometimes fatal even to man.

Flying half-beaks are fish with long, slender upper jaws and practically no lower jaws. They make long glides over the water and may represent an ancestral form of flying fish.

The elephant fish is so-called because of its very rough thick skin and apparent extreme clumsiness of its body, both characteristics of the elephant. Elephantichthys might be likened to a thick leather bag about eight inches long stuffed loosely with vital organs. It has a cartilaginous rather than a bony skeleton. It flattens out when laid on a flat surface out of water. It is almost mollusk likee in the softness of its body. Its skin is approximately a quarter of an inch long.

The aptocyclus, or "rattling fish", is a close relative of Elephantichthys in Arctic waters. It also seems to be a haphazard conglomeration of vital organs stuffed in a bag. The fish actually rattle inside when the skin is not filled with water. All fish of this family live at the bottom of fairly shallow water, firmly attached to flat stones by disk-like suckers. Although they have the power of locomotion they seldom use it, remaining stationary on the bottom and waiting for their food to come to them.

Most fish have a tail fin, usually forked, with which they propel themselves, but the rat fish has a body tapering down to a long, pointed extension that looks like a rodent's tail. They are dwellers in deep waters all over the world. Some are quite fantastic. One, Macruroides inflaticeps, consists essentially of a head and a tail without any apparent intermediate body; it looks like an enormous tadpole.

Pearl fish are minute animals that are sometimes found inside oysters and clams entirely encrusted with mother-of-pearl. They actually become large pearls shaped like fish. These small, nearly transparent creatures sometimes back into the open shell of an oyster or clam that snaps once the fish are inside. When this happens the creature perishes but sets up an irritation that leads to the pearl secretion over it.

LOVE LIFE AMONG THE SPIDERS

There is love and courtship among spiders, as among birds and mammals, but with a unique—and fatal—difference. An observer thus describes a courtship scene in the *Cambridge Natural History*:

"When some inches from her he stood still. She eyed him eagerly, changing her position from time to time. He, raising his whole body on the other side, leaned so far over he was in danger of losing his balance which he only maintained by sidling rapidly toward the lower side. Again and again he circled from side to side, she gazing toward him in a softer mode and evidently admiring the grace of his antics. This was repeated until we had counted 107 circles made by the ardent little male. He approached nearer and nearer and when almost within reach whirled madly around and around her. She joined him in the giddy dance. Again he fell back and resumed his semi-circular motion. She, all excitement, lowered her head and raised her body so that it was almost vertical. Both drew nearer. She moved slowly under him, he crawling over her head. Thus the mating was accomplished.

"A few minutes later, however, the female had eaten her ardent lover."

THE LACE WEAVERS

For 300,000,000 years tiny animals have been weaving delicate lace. They weave constantly, rapidly and in lovely, open mesh patterns. They make a stiff stable lace. Their own limestone entombed bodies are the threads. Night and day, millenium after millenium, they weave and weave, for the curse of weaving is forever upon them. Through time they have covered hundreds of square miles with white and green veils. For the most part these are fragile and short-lived, but in a few cases they have been preserved untorn through the ages.

These lace weavers are the bryozoa, or moss animalcules—one of the oldest, most abundant and least known forms of animal life. They have much the same habits as the corals, but the two limestone secreting creatures are not even remotely related. The weavers are far higher in the scale of evolution than the island builders. Their family associations long have been in dispute. They have been associated with the rotifers and mollusks and even with some unknown ancestral form leading to the vertebrates. Now, however, it is believed that their nearest relatives are the nearly extinct brachiopods, or lampshells.

The two groups started at about the same time in the Cambrian geological period of half a billion years ago, but they followed different paths of development. Both might be considered proto-mollusks—very remotely kin to clams and oysters. For milleniums the brachiopods flourished in the primaeval seas. During the Permean period, about 300,000,000 years ago, they constituted one of the most abundant forms of animal life. Now they seem close to the end of the road. The weavers are as flourishing, and busy, as ever.

Like a coral polyp or the larva of a clam, the bryozoan starts life as an almost invisibly minute, free-swimming creature, usually less than a thirtieth of an inch long. After a few weeks it settles on some hard surface, usually a stone, and secretes its limestone shell. New individuals rise from the body of the founder of the colony at various angles, depending on the particular design of the tapestry being produced. Each of the buds, after achieving its coat, sends out new buds. This is the weaving process.

The outside of the stone coat often is marked with delicate and bizarre designs discernable under a microscope. These designs always are the same for members of a colony and quite similar for an entire species. They make it possible to identify species in geological formations and this eventually may be of considerable importance for oil geologists. After death a colony usually is broken up quickly by wave action. Sea bottom ooze often is filled with the remains. This ooze, over periods of milleniums, becomes compacted into rock.

The weaving process may be very rapid. A colony, starting with a single free-swimming larva, may cover as much as 100 sq. feet. Such colonies have been found on a single stone. They often are found on mollusk shells. At present the bryozoans are economically important chiefly as a menace to the oyster industry. Once they have covered an abandoned shell, oystermen believe, no other oyster will make use of it. About their only other importance to man comes from the fact that some fresh-water species may clog water pipes by their rapid growth.

Every bryozoan in a colony remains throughout its life a separate animal, shut off from its fellows by a wall of limestone and leading an independent existence. Nevertheless, in the species pattern it assumes, each colony acts as if it were a single organism.

Moreover, a phenomenon unique in nature, every individual appears to be two and in some species three animals in one. Each leads its own life and dies its own death at its own time. But all make up a single microscopic whole.

First is the zooecium, a limestone-encrusted box of tissue. This is the continuing individual. Inside the box is a little tentacled worm, the polypide. It contains all the vital organs—the brain and the nerve, circulatory and digestive systems. It breathes, hunts, eats and lives quite independently of the zooecium. This polypide usually is short-lived. It has no excretory system. Poisons pile up. It degenerates and dies. When it expires the cells of the zooecium wall assert themselves. From the dead cells of the polypide they extract what nutritive material is present. The "inside animal" becomes a brown speck-like body. Then the zooecium cells sprout a bud which becomes a new polypide. This lives its normal life span and suffers the same fate as its predecessor. Another brown body is the only evidence that it has lived. This process may be repeated ten or twelve times. Think of a man, or any other high animal, which could replace over and over again its entire internal system with another made out of its own skin which had eaten its own defunct brain and heart.

The relation of zooecium and polypide as it exists in one type of bryozoa, the so-called "sea mats", was vividly described by the great British naturalist P. H. Gosse. These are not lace weavers. They form a colony which looks like a pale, yellow leaf, such as Gosse found in a microscopic study of a mass of sea weed in which he saw other animals like "exquisitely crimson leaves thinner than the thinnest tissue paper, with tall and elegant dark red feathers and purple filaments each as fine as a silk worm's thread."

"Each individual cell [zooecium] of the sea mat", Gosse tell us, "is shaped like a child's cradle. Suppose a coverlet of transparent skin were stretched over each cradle, leaving an opening just over the pillow. Suppose in every cradle there lies a baby with its little knees bent up to the chin in that zig-zag fashion in which children often lie.

"But—the child is moving. A slowly pushed open semi-circular slit of the coverlet and we see him gradually protruding his head and shoulders in an erect position, straightening his knees at the same time. He is raised half out of bed. His head bursts open and becomes a bell of tentacles. This baby is the tenant polypide.

"The chambers themselves show signs of life. Their front doors suddenly open, gape widely and shut with a snap. This opening and shutting is repeated over and over again. The polypide emerges from the cell slowly and withdraws like lightning at the slightest alarm."

As mentioned before, some bryozoans appear to consist of three animals in one. The third is the so-called avicularium, or bird's head, also vividly described by Gosse: "The cells [of this particular species] are oblong-shaped, and look much like a sack of corn. Just below one of the spines that crown the summit of the cell on one of the edges is situated a small lump which bears a remarkable resemblance to the head of a bird. It has a strongly hooked beak with two well-formed mandibles, one of which is removable. You observe it deliberately opening, like the beak of a bird and then closing with a strong, sudden snap. The birds' heads are not inhabitants of the cells. They are not even integral parts of them. The cells have their own proper inhabitants, each leading its own life and each essentially formed on the same plan as that of the baby in the cradle. There is no visible connection between its and the bird's head, which is cut off entirely from the interior of the cell. This head has a muscular system entirely its own. It seizes small animals but has no means of passing them into its mouth".

The real function of these avicularia is unknown. They have been pictured as fierce watchdogs kept by the bryozoa for defense against approaching enemies. Gosse speculates that they may serve indirectly as hunters, seizing and killing small animals. The disintegrating bodies of their prey, attract hordes of smaller sea creatures which can be gathered up by the tentacles of the polypide.

THE WAYS OF CRABS

Crabs that wear clothes, others that carry arms, and still others that march like regiments of soldiers are among the curiosities of Australia's Great Barrier coral reef.

One crab forces the coral polyp to build a limestone palace for its abode. The female of this species lodges on the polyp when it is in the larval state and causes an irritation which forces the host animal to build up the walls. The resulting house is just big enough for the crab to move about in comfortably. There always is a door through which she obtains her food.

Another species merely sits on the end of a sprouting coral which, growing outward, makes a long, circular burrow for the crustacean. Through this it can move backwards and forwards at will. The forward part of its body is enclosed in a hard shell the exact color of the coral so that when the crab sits at the door of its burrow it cannot be distinguished from the coral.

Still another crab carries two sea anemones, one in each "hand", wherever it goes. In its first few months of life it seizes these plant-flowers—living animals with stalks and petals like flowers which ordinarily are attached to rocks under the water—about the centers of the stalks. Thenceforth it moves about like a person carrying two umbrellas.

The most logical explanation of this behavior is that the anemones serve as weapons, killing or paralyzing small sea animals which come in contact with them. This species of anemone has stinging cells in its disk. These curious weapons are carried by the crab continuously and seem essential to its life. When one of them is taken away, the crustacean moves automatically to grasp it again. When a crab is killed slowly in alcohol it clings to its weapons even in its death struggles.

There are spider crabs which cut and wear clothes. They cut off pieces of living sponges and place them on their backs. These sponges become entangled in tiny hairs which protrude through the animal's shell, and continue to grow until they protrude several inches over the back. Thin layers also cover the under part of the body and the legs. Every time a crab sheds its shell, it must make itself a new suit The practice probably is beneficial to both animals. The crab, living in a forest of sponges, looks like a sponge itself and is thus concealed from its enemies. The sponge benefits by being carried to new food sources. When the shell is shed the sponge simply attaches itself to a rock and continues to grow.

One of the most remarkable cases of commensalism in nature has been found by Dr. Melbourne Ward, Australian zoologist in a degenerate type of barnacle which makes its way through the thin shell of one of the Barrier Reef crabs. It wanders through the blood stream of the crab and finally comes to the surface where it forms a little sac for itself. Here it metamorphoses into another form and sends long, thread-like filaments into every part of its host's body. In some respects it is like a cancer among higher animals, except that in this case the malignant growth is that of an individual animal of another species. It lives off the food eaten by the crab but never kills nor apparently seriously injures its host. The one notable effect, for which there is no adequate explanation, is that it changes a male crab into a female.

The soldier crabs are beachdwellers, about two inches long. They march across the hard sand in perfect order, as if they were under the control of leaders. No "officers", however, have been observed. When approached, they burrow rapidly in waves, like a regiment of infantry. First the front rank disappears in the sand, followed in order by those behind. The regiment disappears completely in a very short time.

The soldier crabs can hardly be driven into the water. When Dr. Ward succeeded in pushing a few of them off the shore they were set upon by ferocious small fish which rapidly devoured them. Realization of this danger apparently is instinctive in the animals.

Some of the land-dwelling crabs of the mud flats dig very intricate burrows with labyrinthine cross and side galleries. Some species live in a communal life. Each crab has its own burrow, but from each there is a passage into a large central hall which seems to be a community gathering place. Other species are intensely individualistic. Each excavates an elaborate labyrinth in the mud, considers this its own home, and vigorously defends it.

During courtship some of these mud crabs perform dances like the courtship dances of birds. The male of one variety, after attracting a mate by his dancing, picks her up bodily in one of his nippers and carries her away. Another variety of sand crab seems to have perfected an engineering technique which still evades human skill—that of building a burrow in soft, dry sand. These burrows are about two inches in diameter. The crab is able in some mysterious fashion to compress the soft sand into a solid substance with its nippers.

In precision of instinctive behavior, Dr. Ward found, these Great Barrier crabs come quite close to the spiders, their distant relatives.

TICKS WITH NOSES IN THEIR LEGS

Ticks, remote spider relatives, smell with their front legs. When these legs are amputated the tick shows no reaction to odors. It cannot smell blood but will feed on any sort of liquid sucked through a warm, moist membrane like the skin. Presumably such a tick in nature recognizes an animal as a proper source of food by smell, while a combination of warmth and moisture from the skin gives a stimulus for feeding.

THE FOURTH REALM OF LIFE

There is a wind-tossed green-grey ocean between earth and sky. It is a sea on stilts, the world's fourth realm of life. There are plants and animals of the land, of the water, and of the air—and there are plants and animals of the canopy of the rain forest, a thousand-mile-wide broken belt around the world. It covers several million square miles—the jungles of South America extending northward into southern Mexico, the basins of the Niger and the Congo, strips of southern India and Ceylon, much of New Guinea. Life is rather sparse in the perpetual, drenched twilight of the jungle floor. It is abundant in the treetops, the habitat of fantastic, and still largely unknown, plants, mammals, birds, snakes, toads, frogs and insects. These might be compared to the flora and fauna of an as yet unexplored continent.

Rain forest trees are, in general, tall, straight, and branchless until near their tops, 100 to 150 feet above the ground. There they send out a rich profusion of branches and foliage. This foliage is like a thick, rough, continuous green blanket held up by tall posts, like a net below trapeze performers in a circus tent. The top of the blanket is a place of intense sunshine. Light grows dimmer and dimmer as it penetrates the leaves and the branches. Finally, on the jungle floor, there is only about a fiftieth as much illumination as on the surface of the canopy.

In the canopy four or five kinds of monkeys take the place of man on earth as the most intelligent and adaptive animals. Primates from the beginnings of the race—the weird, squirrel-like animals of the North American dawn age forests fifty million years ago—have been semi-arboreal.

Most abundant in the tree-land are the pretty, playful, curiosity-driven, humanlike spider monkeys who play tag and throw sticks at each other in the lower branches. Best known, although less likely to be seen, are the big, black, Satanic-looking howlers.

Both of these species, in the long process of adapting themselves to high jungle life, have made third hands out of the ends of their tails. With these highly sensitive prehensile organs they not only clutch branches but sometimes carry out rather delicate manipulations.

Weirdest are the black-and-white striped, woolly-furred night monkeys. These little racoon-like creatures live in holes far up in the treetops. They come out only at night and are seldom seen. They have enormous eyes which shine like live coals among the leaves when the light of a flash lamp catches them.

Probably the most dangerous single animal of the canopy is the tamandua, or golden anteater. It is exclusively a treetop creature, about the size of a rabbit, with golden-yellow, soft, silky fur. It lives almost exclusively on termites which it harvests by sticking its long tongue, covered with a sticky saliva, into their nests. A progressive relative of the sloth, it remains motionless apparently for days at a time and is a slow, clumsy climber.

But woe to anything—jaguar, ocelot, big howler monkey, even man—that runs afoul of it. It strikes suddenly and fast with its long, curved scimitar-sharp claws, and always aims at the stomach which it rips open. No other creature will venture near a tamandua, except by accident. Probably it is voiceless, although natives have attributed to the sinister little anteater a peculiarly weird cry heard in the moonlit jungle. This now is believed to be the call of a bird.

Climbing rats are abundant in the jungle top. They feed, for the most part, on fruits. Here also is the abode of pigmy squirrels which cling, heads downward, to the tree trunks with their tails curled over their backs, squirrel fashion. These animals are about five inches long, including the tail whose length is about equal to that of the rest of the body. There is a tiny, climbing mouse with short, broad feet and sharp, curved claws. Bats, mostly small, fruit-eating animals, flutter about in the darkness. Probably there are few of the big dangerous vampires in the high treetops. They fare better on the blood of larger, ground-dwelling creatures such as tapir and peccary.

RUBBER-BAND WORMS THAT STRETCH AND STRETCH

There is a worm ninety feet long. It is the giant of a family of white, red, yellow, green, purple, and violet worms whose habitat ranges from sea bottoms to jungle treetops. The worms shoot poison-tipped harpoons out of their brains. Most can shrink at will to less than a third of their ordinary length. They always shrink when they die. Some can break up into hundreds of fragments, each of which will grow into a complete new worm. They tie themselves into inextricable knots. They build their houses from the slime of their own bodies.

This class is that of the ribbon worms or nemertina. There are about five hundred known species—perhaps as many more are unknown. Still near the bottom of animal life, they represent revolutionary advances from the lowest of worms, the planarians, with which they share many characteristics. They have evolved integrated brains and nervous systems. They have, for the most part, taken on a true worm shape. They have acquired weapons and, in some cases, arsenals of weapons. They have eyes that see. They have a digestive system, a mouth near the front of the body, and closed blood system through which flows a liquid which usually is colorless as water. Perhaps they hear. At the top of the head in certain species there is a group of cells with hairs and bristles which may constitute an organ of taste. Along the way of achieving these advances they have given up a little freedom and a little immortality for a little more efficiency.

Those best-known are inhabitants of sea shores, especially the Atlantic coasts of North America and Europe. They live under rocks, in abandoned mollusk shells, in windrowed masses of sea weed, in thin, parchment-like tubes which they secrete from their own skin. Their general appearance is that of a tangled mass of slimy string, but some members of the family have among the most brilliant color patterns known in nature.

The most conspicuous organ of these primitive worms is the proboscis, a hollow string which is shot out with great speed and force from the front end of the usually cylindrical body. At the end of the string, in several groups, is a sharp-pointed, barbed spear-like stylet with which the prey, usually some minute water animal, is speared. The victim then is drawn back into the mouth by the attached hollow thread. Some groups have no stylets. The thread, upon which is a mucilage-like mucous, is used like a lassoo and coiled tightly around the prey.

The proboscis is associated so closely with the brain that, like the retina of the eye, it has sometimes been considered an extension of it. The thread often is as long as the worm itself. It is shot out with such force that is frequently breaks off and continues to lead an independent life for a few hours. A new proboscis always develops.

When coiled, the proboscis rests in the center of the two-lobed brain. It is continuously shot out and pulled in and probes the water around it. Presumably at first it was an extremely sensitive sensory organ by which the brain was kept aware of its surroundings. The attached stylet, an offensive weapon, was a later development.

In a few cases the thread carries a multitude of unattached barbed points, a sort of machine-gun arrangement, which can be hurled in all directions in the hope of hitting something. It also carries tiny hooks by which it can be attached to some object. By means of the attached line the worm pulls itself forward over beach or sea bottom, its ordinary means of locomotion. It is also able, however, to glide like a planarian and to swim.

Nemertinea breath through the walls of the oesophagus, or gullet. When the tide comes in, shore-dwelling species rapidly swallow and eject mouthfuls of salt water. Oxygen to purify the blood is obtained from the water. The blood circulates in two or three vessels. It is a colorless plasma in which float both green and red corpuscles.

There is little knowledge as to the precise nature of the nemertinean sensory organs. There are, however, nerve cells in all parts of the body and the animal is quick to respond to any irritation, especially to any chemical change in the water. With an intense stimulus the body is contracted violently, twisted, and even torn apart. Even a headless specimen will move toward food placed nearby. A severed head may continue to creep restlessly for several hours. The headless body moves only when stimulated. With most mud-dwelling species it is difficult to secure an entire specimen. The slender, fragile body is likely to break into many fragments when disturbed. Quite commonly, even without any particular disturbance, a large worm will break up into a dozen or more pieces. Each becomes a small, new animal. Some regenerating fragments secrete disks of mucous and form cradles, in which they may remain for months while new organs are being formed. Eventually the disk ruptures and the new worm emerges. There is a specific tendency in some species thus to reproduce during warm weather, with a brief period of sexual reproduction during the cold months.

These worms are extremely tenacious of life. Even without food they may live as long as a year in the proper environment. Ordinarily they are quite voracious animals. They eat earthworms, other sea worms, small mollusks—almost anything soft-bodied which the eternally active proboscis can bring to the mouth. There it is sucked into the digestive tract. The digestive process is very rapid. Some species have distensible mouths. Like snakes, they can devour animals bigger than themselves. Some are cannibals. When times are hard they can, like planarians, absorb themselves. A case has been known where a nemertean digested all but a twentieth of its own body in a few months, apparently without any ill effects. The lost tissues were restored as soon as food again was available.

FROG VERSATILITY

Animals of many talents are the frogs. Some grunt like pigs, others cackle like hens. Some chirp like crickets, others caw like crows. Still others quack like ducks. There are golden frogs, scarlet frogs that play dead, frogs that build houses.

All this assembly is found in one small corner of the world, southeastern Brazil. This particular tropical countryside long has been known for the abundance and variety of its amphibian life.

Some of the frogs in this area are particularly notable for their coloring. Two are almost solid gold in color. Perhaps the most notable is Brachycephalus ephippium, which not only is brilliant gold in hue but has armor plates of bone on back and head, and whose tadpoles are nearly three times the size of the adults. All the adults, less than an inch long, have the armor plate strongly developed, although the shape and size shows considerable variation. The general form of the bony deposition just under the skin, in no way connected with the skeleton, appears to be typically that of an hour glass across the back with one or more separate bony islands. Sometimes these islands are fused with the hour glass. The adults hide under leaves and fallen tree trunks in high mountain woodlands and come out in large numbers only in rainy weather. They appear to be rather clumsy creatures. Their gait is a slow walk.

The nightly chorus of certain of the frogs sounds like a regiment beating on tin pans. Others have calls that are like the sounds made by winding a watch or filing iron. The "tin-pan frog" is one of the most conspicuous creatures of the region. The chorus of singing males gives a booming metallic sound which seems at times to be a regular clanging, like that of a blacksmith hammering on an anvil.

The "tin-pan" frog builds its own house—a crater-like structure of mud projecting above shallow water within which its eggs are laid during the dry season. These nests usually are constructed close to the water's edge. Here the eggs hatch and the young tadpoles are swept into the pond by the next heavy rain. The mud walls apparently protect the eggs from depredations by fish. Adults stay in trees except at the time of egg-laying. The male is said to come to the pond first to build the nest, before the female arrives to lay the eggs. The frog that quacks like a duck is a closely related species. It has a peculiar habit of swarming. Hundreds may appear at one time in a single tree.

One of the golden frogs is about three inches long and almost pure gold in color. Its voice is like the slow grunting of a pig. It sleeps during the day in large leaves of bromeliads, trees of the pineapple family that often hold rainwater in their axils. They sometimes are described as living "tubs of water." At night the frogs come down out of the leaves and go to ponds and streams in the neighborhood in search of insects. Their leaf sleeping chambers apparently give them complete protection from their natural enemies.

One gray and brown Brazilian frog, extremely sluggish by day, when handled assumes a wooden, dead appearance, with the limbs brought close to the body and the head bent forward, so that it resembles a patch of fungus or a chip of wood. Even when left on their backs for a long time they continue to play dead.

A notable singer among the Brazilian tree frogs is Hylabypunctata, whose call is a high, frequently repeated tit-tit-tit. When many sing together the chorus is so loud it can be heard nearly a mile away.

One brilliant-red-legged frog, brought to Washington by the Smithsonian Institution, ate nothing for seven months and did not change its position for days at a time. Throughout this period it seemed to lose no weight. At the end of seven months it eagerly ate worms and files.

A violet frog that lives in the clouds and sings like a bird has been discovered by Dr. Bertha Lutz of the National Museum of Brazil on the summit of 10,000-foot-high Mt. Itatiaia in the Mantiquiera mountains. This frog, hitherto unknown to science, has a purple back spotted with gold, bronze and deep yellow. Below the purple is a deep violet blue.

Since the Mantiquiera mountains, the highest in Brazil, are almost perpetually cloud-veiled, the little animal appears to be entirely a creature of cloudland. Its curious colors perhaps have been borrowed as camouflage from the sky. It has a weak voice and its song is very much like that of a bird. It is found in swift mountain brooks, part of whose courses are subterranean.

THE HORNED VIPER SPEARS OTHER ANIMALS

Best-known Egyptian cobra is the so-called "spitting serpent" or Libyan asp. It supposedly has the ability to spit in the eyes of its enemies, such as dogs, and the saliva temporarily blinds the victims.

The cobra was a sacred animal in ancient Egypt. It was associated with the sun and with royalty. It formed part of the head dress of solar deities and was represented in the crowns of kings and queens. Toward the end of the 20th dynasty, when it became the custom to preserve sacred animals, it was embalmed at Thebes.

There is a fair possibility that one of the sixteen varieties of Egyptian cobras was the "asp" with which Cleopatra took her own life. It is more probable, however, that she used an even weirder and almost as deadly snake, the horned viper. This serpent is common on the fringes of the Egyptian desert. It buries itself in the hot sand, only its eyes and the top of its head being visible. Its two horns resemble barley seed and attract birds within its reach. When disturbed it can throw itself forward. It was called "aculum" (spear) by the Romans because of this darting motion.

THE WORLD OF INSECTS

The Roman naturalist Pliny wrote of ants in the Himalayas "the color of a cat and as large as an Egyptian wolf." Pliny naively had accepted tales of travellers but the actual curiosities of the insect world are almost as strange as anything he related. There are bugs that live in ice, bugs that are happy only in near boiling water, snow white bugs that dwell deep in the earth, bugs that make their homes in petroleum pools.

None are as big as wolves, but the insect world has its giants as well as its dwarfs. The Atlas moth of India has a wing-spread of nearly a foot. An East Indian walking stick is 15 inches long. The Hercules beetle of Africa sounds like an airplane in flight. Enormous forelegs, more than twice the length of the rest of the body are characteristic of a black wood beetle which covers a space of eight inches with all its legs extended. A curiosity of the Malay Archipelago is a "fly with horns." It has protuberances on its head which suggest the horns of a deer.

A South African fly has eyes which extend on stalks from the sides of its head. The stalks are so long that the measurement from eye to eye is a third more than the length of the body from head to tail.

One blood-sucking insect can distend itself with blood to more than twelve times its original weight. As the huge meal is digested the abdomen contracts like a deflating balloon.

The death watch beetle, standby for stories of haunted old castles, bumps its head on the top of its tunnels in wooden walls to send a kind of telegraphic message to its mate.

Some chalcid flies paralyze caterpillars and lay self-multiplying eggs in their bodies. More than 2,000 larvae may be produced from a single egg deposited in this way.

A singular ant lion, dweller near the Egyptian pyramids, has a slender and elongated neck whose caliper jaws seem to be held at the end of an outstretched arm. The neck, in many cases is far longer than the rest of the body. It permits the insects to probe for prey in deep crevasses.

The goat of the insect world, the drugstore beetle, is known to consume 45 different substances, including the poisons aconite and belladonna. Other beetles feed on cigarettes, mustard plasters and red pepper. Ants have shown themselves resistant to cyanide. In the case of some insects a reduced diet slows down growth. Some wood-boring grubs sometimes live in house timbers for years after they have been put in place. In one instance an adult beetle emerged from a porch post that had been standing for twenty years. The dried timber lacks the nutritive qualities of the living tree and the growth of the grub is arrested so that long periods pass before it reaches maturity.

A carnivorous butterfly larva lives in the nests of an Australian ant where it feeds on the young. An especially tough outer shell protects it from attacks by adults ants.

The rat-tailed maggot inhabits stagnant water. It feeds on the bottom and breathes air through an extensible tube that forms its tail. Like a diver obtaining oxygen through an air hose while working on sea bottom, it is able to remain submerged as long as it desires.

The little frog hopper produces its own climate. In spring and summer small masses of froth often appear on grass stems and weeds. Within such a bubble mass, sheltered from direct rays of the sun and kept moist by the foam, the immature insect spends its early days. For millions of years it has been employing its own primitive form of air conditioning.

GIGANTIC SERPENTS OF THE SKY

Titanic pink serpents coiled and wheeled in the sky. The earth below was plunged in a chill twilight as they shut out the December sun. These cosmic reptiles were two or three miles long. They moved about a mile a minute. They made a noise like a tornado punctuated with the rat-tat-tat of machine guns.

Thus the naturalist John Audubon described a mass passenger pigeon flight over Kentucky which, he estimated, included more than a billion birds. As they came out of the northeast they looked like a gigantic, low pink cloud driven by a hurricane. Suddenly they split with almost military precision into the coiling, snake-like formation as predacious hawks hovered above them.

When these hawks came, says Audubon, at once with a noise like thunder they rushed into compact masses, pressing upon each other towards the center. In these almost solid masses they darted forward in undulating lines, descended and swept close over the earth with inconceivable velocity, mounted perpendicularly so as to resemble a vast column, and when high were seen wheeling and twisting in continuous lines which resembled the coils of gigantic serpents.

When the birds reassembled from their emergency snake formations, they constituted, Audubon estimated, a column one mile broad passing overhead at the rate of a mile a minute for three hours. Thus the solid mass of the birds would have covered 80 square miles. Such a monster would have required, the naturalist calculated, about nine million bushels of food a day.

It is more than a century since anybody has witnessed such a phenomenon. Civilization and nature combined to destroy the almost incalculably vast hordes of pink-breasted birds which, acting in a weird unison, seemed to the pioneers like cosmic monsters invading the earth. Hundreds of millions were slaughtered by hunters. Millions perished in one great Atlantic storm when, it was reported, the sea over a radius of three or four miles was covered completely with their bodies.

The passenger pigeon long has been extinct. The last survivor of the tornado-like masses now is mounted and on exhibition at the Smithsonian Institution. It died in captivity in the Cincinnati Zoological Park at 1 p.m., September 1, 1914. Every year Smithsonian ornithologists get reports that one of these birds has been seen in some remote forest. Almost beyond question, however, these reports are due to the wish fulfillment of amateur bird watchers.

The extant mourning dove sometimes is mistaken for the passenger pigeon. In the west the band-tailed pigeon has been similarly mistaken. Even expert ornithologists might make such errors from casual observations. Although convinced that the bird is extinct scientists continue to investigate any plausible clue to its survival.

According to Smithsonian Institution ornithologists, there is a popular idea that the passenger pigeon mysteriously disappeared and that, while still enormously numerous, it suddenly ceased to exist. Its annihilation has been attributed popularly to various natural phenomena and it has even been rumored that the bird migrated to South America. The natural phenomena supposed to have been causative of its extinction are epidemics, tornadoes, early deep snowstorms, forest fires, strong winds while the birds were crossing large bodies of water which caused exhaustion and death by drowning. Circumstantial reports were published of immense numbers drowned in the Gulf of Mexico, a region well beyond the usual range of the bird. Destruction of the forests undoubtedly was a large detrimental factor in the life history of the pigeons, for the forests supplied their principal food as well as roosting and nesting places.

A bird accustomed for ages to living together in large numbers and close ranks, whether in feeding, migrating, roosting or nesting, might find it impossible to continue these functions with greatly reduced and scattered ranks. It is probably more than a figure of speech to say that under these circumstances such a communist bird would lose heart, nor is it fanciful to suppose that sterility might in consequence affect the remnants. Our continent is so well known that accounts of the presence of living birds must be considered more than doubtful.

The mass flights came about once every ten years in the early winter. The normal habitat of the pigeons was in the great forests of Quebec and Ontario. There they were widely scattered, feeding chiefly on acorns. When snow covered the ground they moved southward, but ordinarily not in great masses. But a periodic failure of the acorn crop, of the extent of which the birds seemed to have some mysterious awareness, caused them to assemble in one body and start a mass migration southward, obscuring the sun for hours as they passed beneath it.

Like tornadoes, they wrecked forests in their flights. Says the naturalist Alexander Wilson: "The roosting places sometimes occupy a large extent of forests. When they have frequented one of these places for some time the appearance it exhibits is surprising. The ground is covered to a depth of several inches with their dung. All the tender grass and under wood is destroyed. The surface is strewn with large limbs of trees, broken down by the weight of birds collecting one above the other. The trees themselves for thousands of acres are killed as if girdled with an axe. The marks of the desolation remain for many years on the spot. Numerous places could be pointed out where, for several years after, scarcely a single vegetable made its appearance."

After these mass migrations from the north the pigeons scattered through the forests in search of food but assembled again in the spring for egg-laying and hatching. Wilson reported: "Not far from Shelbyville, Kentucky about five years ago, there was one of these breeding places which stretched through the woods in a north and south direction several miles in breadth and was said to be more than 40 miles in length. In this tract almost every tree was furnished with nests wherever the branches would accommodate them.

"As soon as the young were fully grown numerous parties of inhabitants from all parts of the adjacent country came with wagons, axes, beds and cooking utensils, many of them accompanied by the greater part of their families, and encamped for several days at this immense nursery. The noise was so great as to terrify their horses and it was difficult for one person to hear another speak. The ground was strewn with broken limbs of trees, eggs and young squab pigeon which had been precipitated from above and upon which herds of hogs were fattening. The view through the woods presented a perpetual tumult of crowding and falling multitudes of pigeons, their wings roaring like thunder, mingled with the frequent crash of falling timber."

The last great nesting was recorded at Petoskey, Michigan, in 1878. The area covered is said to have been forty miles long and 30 miles broad.

Systematic commercial hunting of the birds reached its height shortly after the Civil War. In 1879 dead birds were sold on the Chicago market at 50 cents a dozen. Pigeon hunters made from $10 to $40 a day.

THE LIMBLESS LIZARD

A supposedly welcome guest in the underground chambers of leaf cutter ants is the amphisbaena, a nearly limbless lizard about a foot long which looks something like a gigantic earth worm. This creature, seldom seen, ranges from northern Brazil to lower California. When out of its habitat the amphisbaena is almost helpless and moves along the ground with feeble wriggles. Some species lay eggs; other give birth to living young.

THE MADDENING TARANTULA

The tarantula of southern Europe—a large, hairy spider—long was credited with causing a weird, infectious madness by its bite.

The first reported effect of its poison—actually quite mild—is said to have been to put the victim into a deep lethargy from which he could be roused only by music which set into motion an overpowering impulse to get up and dance. Once the victim started to dance he could not stop until he fell to the ground from exhaustion. Then the condition supposedly was cured for a year. On the anniversary of the bite, however, the dance was involuntarily repeated. From the tarantula's first victim the dancing mania allegedly spread like a contagious disease through the surrounding countryside. The name still is used both for an Italian dance and for the music which accompanies it.

The tarantula is a subterranean creature which hibernates in its burrow during the winter. Bees and wasps are said to be killed almost instantly by its bite. The spider always strikes at the junction of the head and thorax.

A FLOWER THAT GROWS THROUGH SOLID ICE

A plant that drills through several inches of solid ice to bloom in early spring is the blue moonwort of the Swiss Alps. It belongs to the primrose family. In autumn it develops thick, leathery leaves. These lie flat on the ground, expectant of the snow and ice sheet that may cover them to a depth of several feet.

When spring arrives and the hot sun melts most of the snow and some of the ice, water trickles down to the rootlets and arouses growth in the sleeping plant. Internal combustion ensues with the floral tissues. The resulting heat melts the ice about the uprising flower buds and the stem pushes its way upward. More water flows to the roots and finally the plant tunnels a passage to the air and sunshine. So long as the heat given off from the growing stem and buds is sufficient to prevent solid freezing of the parts the plant is indifferent to the surrounding ice cold temperature. It undergoes the usual transformations, is fertilized by early bees and forms many hundreds of wonderful blue flower groups which look as if they were beds over a thick layer of transparent ice. The leaves are now no longer thick and fleshy, but thin and papery. They yield up their carbon compounds as fuel to melt a tunnel through the ice and production of buds and blossoms on a flower stem above the ice mantle.

THE VERSATILE ANT FARMERS

There are microscopic "farmers" whose fields are measured in fractions of inches. They are ants—the most widespread fungus-growers in the Western Hemisphere. Their range extends from Florida to Brazil. They are tiny creatures, seldom noticed, who cultivate a species of yeast which is their sole food.

The ways of life of this curious ant with the formidable scientific name of cyphomyrmex rimosus minutus, have been studied throughout their habitat by Dr. Neal A. Weber of Swarthmore College.

"The ant," says Dr. Weber, "is versatile in the American tropics where the humidity is high and the temperatures uniform. The most common sites are in clay soil on the forest floor. An empty snail shell, a curled dead leaf or a rotted twig may suffice for a colony of these small ants or they may find requisite conditions among roots or in the dead wood high in the rain forest canopy.

"During the rainy season in Panama City there was a nest on a concrete cylinder above ground which protected a gas meter. The cylinder was 17 centimeters high (about 6 inches), by 36 centimeters in diameter and was covered loosely by a concrete cover. In the narrow space on the rim under the cover a colony had walled off an elliptical area 36 by 17 millimeters (about 4 inches by 3/4 of an inch), in which the entire nest with a fungus garden was formed. During drier periods the ants would move down into the soil.

"The workers usually are slow-moving and become immobile at the slightest disturbance. Sometimes, however, they run as rapidly as the average ant when disturbed and seek to escape rather than feign death. In "feigning death" the ants quickly curl up their legs and fold their antennae close to the body so that they appear almost invisible bits of dirt when casually examined.

"The ants spend much time in grooming the forelimbs, antennae and other parts of the body. Regardless of how dusty an ant may become momentarily, it keeps its antenna immaculate by drawing it through its mouth and licking and cleansing it. They also clean one another. In grooming each other the ants may carefully go over a large portion of the body. In one instance a slightly callow worker was watched as it groomed another of the same age. The one being groomed turned over on its side, like a dog or a monkey. The grooming of each other and the cleaning of the brood is a vital part of their activities as it removes alien bacteria and fungi and also may have a nutritive function so far as the brood is concerned.

"The fungus garden consist of masses from a quarter millimeter to a half millimeter in diameter (from about 100th to a 60th of an inch.)"

They have their bitter, nearly microscopic enemies. Upon them, as upon elephants, ride much smaller, bareback riding mites whose acrobatic stunts would be the envy of any circus performer.

"Seven out of 16 ants so examined," Dr. Weber says, "had mites on them. These mites have no difficulty in moving from one site to another on the ants. A transfer of a mite from one ant to another was watched. It had been riding on one ant when another brushed by waving its antennae over the other as is customary. In a flash the mite grabbed the tip of the left antenna. The ant did not attempt to dislodge the mite although it already had two others on its body. The mite had a rough ride, but was not dislodged."

The peculiar type of fungus grown by the ant does not grow naturally outside the nest. It can be isolated and cultivated but it quickly is overwhelmed by other fungi in any artificial culture. It is probable that ant and fungi need each other for survival. Possibly the saliva of the insect is essential for the growth of the primitive plant. Likewise the peculiarly developed fungus is essential for the well-being, even for the survival, of the ants. It is one of nature's partnerships.

OSTRACODERMS: ANCESTORS OF TRUE FISH

The race of fish first appeared about 350,000,000 million years ago in the Silurian geological era. It was made up of grotesque, clumsy, heavily armored animals who crawled over the ooze of the sea bottoms with very little, if any, capacity to rise or propel themselves in the water. The ascent from such an unpropitious beginning to the swift, graceful swimmers of today is one of the wonder stories of evolution.

These Silurian animals were the ostracoderms. They belonged to the general fish complex but were not in the direct ancestral line of any extant fish. This race continued, in various groupings, for at least 150,000,000 years. The earliest forms were wormlike animals whose fossils are found in ancient rocks of Esthonia. Their heads and the forward parts of their bodies were covered with bony plates. They had no fins to serve for steering and balancing. In appearance they were close to tadpoles. It is quite obvious that they were bottom-dwelling forms who swam, if at all, awkwardly and laboriously. The evolution into more and more efficient swimming animals can be traced through later and later fossils throughout the life history of the race. The body became more flexible. There was a gradual reduction in the thickness of the external armor as the ostracoderms came to depend more and more on speed and less on invulnerability. At the end they probably were comparatively good swimmers.

A little later than the earliest of this long extinct family came the first representatives of the true fish—probably derived from the same general ancestral stock. They also were bottom-dwelling animals, although from the beginning they appear to have been a little better adapted for swimming. In these also, the head and forward part of the body were encased in heavy armor. In ostracoderms, however, this had formed a continuous shell, allowing no anterior freedom of motion in the water. In the earliest true fish it was divided into two parts, the head shield and the body shield. For the most part, however, they could use only the tail and posterior part of the body for propulsion. But through many generations various diversifications of the race became more and more fishlike in form, shed their heavy protective plates, developed paired fins for steering and balance, and continuously improved as swimmers.

"We must take it for granted," explains Prof. Anatol Heintz, Norwegian paleontologist, "that the ancestral forms of the vertebrates evolved in water. Most primitive forms lived on the bottom and had not yet specialized sufficiently to be able to swim. If the oldest vertebrates were bottom-living or burrowing forms they must have learned to swim, just as later they learned to crawl, walk, run and finally fly."

Among the earliest groups of true fish were the coelacanths, or "hollow spines." They left many fossil remains over a period of 200,000,000 years. Supposedly they became extinct about sixty million years ago, at the start of the dawn age when most higher life types known at present first appeared. Through all the vast eons of their existence the "hollow spines" changed little.

Three years ago came one of the outstanding events in present day biology. A living coelacanth was caught by native fishermen off the northeastern coast of Madagascar. It was quite similar to its fossil ancestors—armored head and all. Apparently the Madagascan fishermen had been capturing similar creatures in their nets occasionally for years, without realizing that they were of any particular significance.

To biologists the news of this capture was as exciting as would have been that of finding a living dinosaur. The coelacanths, in fact were hoary with age when the earliest dinosaurs appeared on earth. This fish was a survivor from days when animals first were developing spines and brains.

The specimen, however, was practically ruined before it came to the attention of the scientists. Native sailors had sliced it open from snout to tail. All the brain and other soft parts of the head were gone. Other parts were so badly mangled that it was impossible to reconstruct them.

Since then several others have been caught. An intriguing possibility is that of obtaining a female with unborn young. A developing embryo supposedly recapitulates ancestral forms. If one could be found it would be possible to reconstruct something of the real ancestry of the first back-boned animals.

Natives report that the coelacanth is extremely oily. Its flesh drips oil. When boiled it quickly turns to jelly. This fact may have a bearing on the origin of some of the earth's great oil deposits. Man today may be running his automobiles or heating his homes on the fuel produced by vast hordes of these head-armored, hollow-spined fish in the ancient warm seas.

THE EVER FAITHFUL HORNBILLS

Lady hornbills are trusting wives and gentlemen hornbills are unbelievably faithful husbands.

The hornbills are birds with enormous beaks. They have the size of small turkeys and are usually found in pairs in the forests of East Africa. They are perhaps best known from the curious instinctive behavior of the female. Before laying her annual quota of two eggs she walls herself with mud, collected by the male, into a hole near the top of some high jungle tree. There one of the eggs—apparently seldom both—is hatched and the chick reared. The female continues this voluntary imprisonment for two months or more.

There is always a small aperture in the wall. Through this the foraging male passes food to his imprisoned mate, once an hour or less. Food consists mostly of fruits. Sometimes he brings her what apparently are playthings to relieve the monotony of hatching and chick-rearing.

A comprehensive report on the behavior of these grotesque birds in the Mpanga Research Forest of Uganda, by Dr. Lawrence Kilham of Bethesda, Maryland, is a classic on bird-watching.

Hornbills mate for life and apparently their conjugal life is a model of high morality for the whole animal kingdom. Walled into the tree-holes, the females obviously are helpless to protect themselves against any infidelity, and, sad to say, there are vampire female hornbills in the jungle whose only thought is to steal some imprisoned lady's spouse.

In the case observed by Dr. Kilham, however, the male preserved his virtue to the end. "By November 8," he records, "the female was walled in, and a more serious attempt at interference was now made by a foreign female.... She was following the male and lighted in the next tree when he lighted above his nest hole. On November 23 the same course of events took place, except that the male was less tolerant. He fed his own mate, then drove the intruder away. A week later I saw her fly in close behind the male and light 25 feet from the nest hole. The male gave his mate a piece of bark followed by some fruit, and then bounced from one branch to another toward the foreign female."

The poor fellow was falling, falling, but "the female within the nest screamed a number of times. I wondered whether the interloper could seduce the male, but from subsequent observations it seemed unlikely that she would. The male returned again to the nest hole, and a few minutes later was in the upper part of the tree knocking about on dead branches until he dislodged a piece of bark. He clamped his bill on the bark until it was largely fragmented. Then he moved toward the foreign female. If he presented the bark [a cherished play object among hornbills] one would suppose that she had some attraction for him. After a moment, however, he changed his direction, flew down to the big limb below, bent over the nest hole, and gave the token to his mate, accompanied by a feeding chuckle. Subsequently he returned to perch quietly within eight feet of the intruding female. At 7:30 a.m. the two of them flew away together. As the nesting season progressed, he became less tolerant of her intrusions...On February 3 I again watched her fly in behind the male and alight on the nest tree, making considerable noise. The male stopped feeding his mate, swooped at the interloper and drove her down toward the ground. However, when he flew away, she followed a short distance behind."

The vampire was hard to discourage. A few days later she was observed at the entrance to the nest, trying to break the wall with her beak. Probably there was a sex murder case in the making. But "After five minutes the male arrived and...drove the foreign female to another tree, flying at her so hard that he knocked leaves from intervening branches. He returned to his nest with a small stick held like a cigar. His mate, who had remained silent, now began her wailing screeches....The intruding female, persistent as usual...had followed the male back to the nest tree. In a few minutes he flew at her again, flying faster than hornbills usually do as he chased her from one tree to another."

But his ordeal of bachelorhood was nearly over. Five days later mother and young emerged from the nest: "The pair of hornbills were perched side by side on their tree. Not long after I heard a great flutter of wings. I looked back to see both members of the pair pursuing a foreign female....When the parents later came to our garden, she did not follow."

ANTS WITH TAILOR SKILLS

Ants developed the craft of sewing long before humans. There are species of tailor ants in Australia, Africa and India that have distinctly ingenious habits. They make nests of leaves sewed together with silken threads, secreted by their own larvae, which they use both as needles and shuttles.

When the nest is torn in any way certain soldiers and workers, apparently specialized for this particular job, rush to the scene. The soldiers arrange themselves to protect the workers. These first try to pull the two edges of the rent together. If the gap is too wide for a single insect to reach the other side and secure it with her mandibles a living chain is formed, sometimes as much as six ants long. One holds another in front of her with her mandibles, the second similarly holds a third, and so on until the other side is reached. Hours sometimes are required before the edges of the tear can be brought together and held in contact.

Then several other workers appear, each carrying a larva head upwards. These little worms are carried back and forth like a shuttle, spinning the threads which are pushed through needle holes made by the workers until the rent is securely patched.

FIEND SYMPHONIES OF THE JUNGLE

Out of green jungle depths at sunrise rises the choral hymn of the damned. It is a symphony of earth's evil, of ancient dinosaurs and flying reptiles, of vampires and witches. It comes from the throats of jet-black, long-bearded, fiend-like creatures wearing red shawls. They are the howler monkeys.

The world's loudest-mouthed bluffers and braggarts are these dwellers in the high treetops. They swear in an ancient tongue evolved over centuries for the effective cursing of hovering white hawks, black vultures and lurking wild cats. Now they curse, loudly and most profanely, airplanes which sweep low over Panama and Costa Rican jungles. They have not found it necessary to invent any new expressions to convey their contempt for the new monsters of the skies.

Their voices are their only weapons. These have proved quite effective throughout the lifetime of the race. The howlers have been able to threaten their enemies with perdition so convincingly that these enemies have believed the threats. Largely as a result, the big black monkeys have been left alone as the dominant animals of the weird, perilous green world at the top of the jungle. They never have had to fight with fists, claws or teeth. All they have done—all it has been necessary to do—is talk about it.

The scream of the howler, hurled defiantly at a possible enemy or raised in a diapason to the sunrise or in a ritual of worship to the full moon, is the most fearsome sound of the jungle. As one zoologist has said: "It's a combination of the bark of a dog and the bray of a mule magnified a thousand-fold." It can be heard, and clearly discriminated, eight or ten miles away. Some say that the howl not only sounds like the voices of fiends let loose from the pits of Hades, but that the appearance of the animals themselves is just about what one would picture for the infernal beings. The loudness and carrying power is due to the monkey's peculiar throat structure, which enables the sound to reverberate. This throat structure is the weapon which nature has provided for the animal and it has enabled him to more than hold his own in the endless struggle for survival of the fittest. Even more, it has made him supremely contemptuous of all lesser-voiced creatures, such as men on foot or men in airplanes at whom he howls defiantly.

Of all apes or monkeys, the howler probably looks the least like his distant cousin, man. He is at very best a grotesque caricature of a chimpanzee or a gorilla. Attempts have been made to oust him from the monkey race altogether and to degrade him to the pseudo-monkeys, the lemurs. But in biology there is nothing to justify this.

Fortunately for students of animal behavior the howler is a daylight animal. He usually goes to bed at sundown and stays there until sunrise, except on occasions when the full moon awakens him and arouses some uncontrollable frenzy which finds expression in the weird howling. So about everything he does is open to observation.

The creatures remain about the least acceptable of the monkey and ape race in human company. The feeling apparently is reciprocal. The howler is an almost untamable wild animal. He never will dance at the end of a hurdy-gurdy grinder's leash, and seldom will be on exhibit in zoos. He dies quickly in captivity, but only after becoming such a nuisance with the howling of a broken heart that zoo keepers are glad to be rid of him. Only one specimen has been kept in captivity at Barro, Colorado—a baby rescued by one of the Indian guides after she had fallen out of a tree. This happens not infrequently to the little howlers before they have mastered the acrobatics of the forest canopy. They are not climbers at birth, any more than seals are able to swim.

In the strange treetop realm among his own the howler is a much more engaging personality than he appears down below. He is the "man" of the green canopy 100 feet above the earth. He is the dominant creature, intellectually if not always physically, and he appears to have evolved a complex form of social organization.

From two to three hundred of the big black monkeys inhabit Barro, Colorado. They are split into groups of from ten to twenty individuals. These groups are probably extended families, each consisting of two or three adult males, a few younger males, and the remainder females and babies. Each clan possesses an area of from 250 to 500 acres. This is the "home town" and few of the monkeys ever stray across its borders.

Within such an area are "roads," path of long branches and heavy vines by which a troop can pass easily from one treetop to another. These same ways are maintained year after year. The howler requires solid footing. Despite his lofty, wind-tossed habitat he is not much of a gymnast. For one reason, his body is too heavy. He appears quite clumsy compared with his lighter, more volatile relatives, the spider monkeys of the same high realm. Howlers, for example, very seldom have been observed leaping from tree to tree. Occasionally, probably only in cases of dire necessity, a swinging vine may be used as a trapeze. Any aerial acrobatics, however, appear far from this monkey's ideas of good sport.

Through its allotted area a group usually moves in single file, the adult males leading the way and the females with young clinging to their backs or breasts bringing up the rear. The treetop roads seldom are wide enough to permit two monkeys to move abreast. When any of the troop drops behind, the procession is held up to wait for him. If he does not appear in a few minutes scouts are sent back to find out what has happened. About the worst to be anticipated is that a mother has dropped her baby. She immediately will descend to retrieve it from the ground or, as is more likely, from some of the lower branches which have broken its fall.

The animals appear to maintain a communistic family life. A family never seems to increase or decrease in numbers. Probably new groups are formed if the birth rate becomes greater than is necessary for replacements. In the absence of epidemics death rates are not heavy, for the animal has no very formidable natural enemies. Its hellish howl is enough to scare away even the strongest, fiercest invaders of its high country.

Classes are mutually exclusive. But there are no wars in the treetops. When one group ventures near the border of a range claimed by another all the inhabitants get together and set up the most fiendish howling of which they are capable. The potential invaders stop and howl back, just as fiendishly. After a more or less prolonged session of this bloodless warfare both factions call it a day and go their peaceful ways. Any actual fight between howler gangs has not been reported by reliable witnesses.

TYRANTS OF THE POLYCHAETE RACE

Knight-warriors and Amazons of the worm world are the aphroditids. They are the aristocrats and tyrants of the polychaete race.

Like the oriental Aphrodite whose name they bear—she was the mythical goddess of love and war who rose from the sea foam armed with golden spears which were the rays of the moon and sun she personified—they crawl over the beach sands resplendent in a bristling panoply of gold and green. Heavily armed for both offense and defense, their prey are all living things remotely their equals in size and strength.

For their battles they carry on their feet "an armory of harpoons, bayonets, lances, spears and billing hooks," says the Rev. George Johnston in his catalogue of annelid worms in the British Museum. "Were it desirable to have any additions to man's weapons of war," he comments, "the aphrodite bayonet might furnish a model for a new kind as formidable as any we possess. It is armed with a kind of pricker affixed to the end of a musket. This appendage is very sharp, formed with several cutting surfaces, and with a spine below pointed backwards which gives it the properties and advantage of a harpoon. Hence, having been forced to penetrate the flesh, the point cannot be withdrawn, but is detached at once.

"This, however, is not the most curious part of the instrument. The bayonet part of the bristle is, in fact, a sheath which encloses another weapon that is exposed only when the scabbard is lost. When we detach the bayonet from the sheath, at the same time we force from its interior a horny stylette with a needle-like point ready to become a good defensive weapon."

The terror of tidal beaches described by Dr. Johnston is the "sea mouse," Aphrodite aculeata, an oval-shaped worm from six to eight inches long and two or three wide. It has from 30 to 50 large "feet" on each side of its body, each carrying an immense tuft of silky green and golden bristles and spines. Many have commented on the malevolent creature's beauty and capacity for inspiring terror.

"The very brilliant iridescent hues," Dr. Johnston says, "are not equalled by the colors of the most brilliant butterflies." "It does not yield in brilliance to the plumage of humming birds or even to the most shining gems," wrote the great French naturalist Baron Cuvier, credited with the original description of the animal.

Normally it moves by jet propulsion. As it goes forward, a current of water is projected with considerable force at short intervals from its rear end. Progress ordinarily is slow, but the sea mouse is capable of considerable speed when pursuing a slow-moving prey. It frequently can be observed motionless, watching a weaker worm or mollusk upon which it is prepared quickly to pounce at a favorable opportunity.

Some of these animals, Dr. Johnston observes, "have 500 feet on each side of the body. Each foot has two branches and each branch at least one spine and one brush of bristles. Thus an individual has at least 1,000 spines. If we reckon ten bristles to each brush, it has at least 10,000."

The bristles, presumably, are almost entirely for defense; the spines for offense, and admirably fashioned for killing weaker animals. Both types of weapons can be retracted entirely inside the foot when not in use, but thrust out again immediately when needed.

Aphrodite hermione, a close relative of the sea mouse, Dr. Johnston points out, "has in the dorsal branch of its feet bristles which may be described as lances. They are so small that a magnifying glass is needed to discover the workmanship, which excels in finish the finest instrument of man by the skill of the most expert artificer. A great number of these bristles garnish the extremity of each foot, and as they are stiff and serially arranged they form a hedge of spears around the body of the worm, placing it within a square of pointed pikes threatening at all points. Other bristles terminate in a knob within which is a barbed lance."

Still others are likened by Dr. Johnston to harpoons, produced from the body only as required. They are very sharply pointed bristles with the point attached to a shaft. The harpoon point, like the bayonet previously described, has a reverted tooth which cannot be withdrawn once it has been plunged into the body of the enemy. It can, however, be detached and left to fester in the wound. Some worms lose all their harpoons in their many fights.

"There is scarcely a single weapon invented by the murderous genius of man," commented the French naturalist Quatrefages concerning aphroditids on Bay of Biscay coasts, "whose counterpart and model could not be found among these worms. Here are the curved blades whose points present a double and prolonged cutting surface, sometimes on the concave edge as in the yataghan of the Arabs, sometimes on the convex border as in the oriental scimitar. We meet with weapons of offense and defense which remind us of the broad sword of our cuirassiers; the sabre-poignard of the artilleryman; the sabre-baionette of the chausseurs. We have harpoons, fishhooks, cutting blades in every form attached to the extremities of sharp handles. Destined to live by rapine and exposed to a hundred enemies, they need such weapons both for attacking and defense."

Some aphroditids swim with ease. The majority, however, are found between tide marks where they burrow in wet sand. A few occasionally trespass in tidal rivers. When placed in fresh water the animals soon die, in their death throes first ejecting a milky-white fluid which turns to blackish-green at the moment of death. Despite their heavy armament, the aphroditids are a favorite food of codfish. They are distributed generally all over the world. The monster of the race in the South Pacific sometimes reaches a length of five feet.

EATING HABITS OF SPIDERS

Spiders digest most of their food before eating. They must subsist on a liquid diet. A powerful digestive fluid from the stomach is discharged on the prey. This completely liquifies the soft tissues. So potent is this fluid that spiders sometimes can devour small back-boned animals, such as fish and lizards, which they kill with their poison fangs. One African species can liquify almost completely a fish two inches long in less than three hours. Another has been observed in captivity to dispose of small snakes in the same way.

THE SUICIDE INSTINCT OF IGUANAS

Some iguanas seem to have the ability to commit suicide without any visible means. Some of these lizards, hitherto unknown to science, captured alive and uninjured in Cuba by Dr. Paul Bartsch of the Smithsonian

Institution, died a few minutes later as if a mere wish to end their lives were sufficient to achieve death.

"These iguanas are vegetable feeders," Dr. Bartsch recorded in his field notes. "They are fairly tame and persisted in chasing the nooses on the ends of our sticks, instead of running their heads through them or letting us place them around their necks. When hard-pressed they finally dash into holes that look like huge crab burrows. When near the coast, where there is a hurricane rampart, they seek refuge in crevices of the rocks. We were surprised when we took those we had captured from our bag on board ship to find four of them dead. Evidently they have a way of ending their own lives."

On Petite Gonave Island off the coast of Haiti are large iguanas which—native fishermen say—can be captured safely only by getting them drunk. Travellers are warned that they are extremely dangerous animals when sober. The fishermen pour rum into hollows of rocks along the shore. The big lizards appear to be very fond of this beverage and drink themselves helpless.

FORESTS THAT EAT MEAT

Relic groves of the great meat-eating forests of 150,000,000 years ago still thrive on the floors of deep, warm seas.

These are made up of plant-animals—predacious trees with red blood and hearts—the crinoids. There are about 700 extant, compared to more than a thousand extinct, species. For a hundred million years they were among the ocean's dominant life forms. Fossil crinoids, or "stone lilies," make up great marble beds in both American and Europe. In 1934 the Smithsonian Johnson expedition dredged nineteen species, including two not hitherto known to science, from the bottom of the great Porto Rico Deep.

The crinoids are highly developed animals, although they look like plants. They can by no means be considered as a form of life on the dividing line of the animal and vegetable worlds. Rather they are animals which have taken on the superficial appearance of plants. They are very highly specialized animals—so much so that there are few places in the world where they can survive in great numbers.

In life they usually are brilliantly colored. Judging from those that are found on the sea bottoms today one of the ancient meat-eating forests must have presented a very colorful spectacle of red, green, purple and yellow "blossoms."

Most of them live in deep water. There are free-moving varieties as well as those that are fixed to the bottom with stems like plants. Until recent years few were recovered in good condition because of the tendency of one of these plant-animals to break itself to pieces when agitated. When brought up from the bottom to the deck of a ship the crinoid would proceed to break off the featherlike arms which make up the blossoms. This was its natural defense reaction in the depths. Its way of escape when one of its arms was seized by a fish was to break it off. Then it could grow another quite easily. As a matter of fact, this is the way the crinoid grows— one of the most wasteful processes of growth in nature. It breaks off one arm and grows two instead; but it cannot increase the number of its arms without discarding an old one.

Another difficulty is that the gorgeous colors of the meat-eating flowers are fast only in salt water. They fade rapidly in air, fresh water or alcohol so that there can be only a fleeting impression of the true coloration.

These crinoids live, for the most part, on diatoms, small crustaceans, and other tiny sea creatures which they first paralyze with poison from the tentacles which line the grooves of the arms through which food is carried to the mouth.

CAVE-DWELLING BIRDS

True creature of night is the guacharo, or "oil bird", of northern South America. It is reddish-brown, about the size of a barnyard hen. Excessive layers of fat built up about its abdomen formerly were valued highly by natives for eating purposes, resulting in the slaughter of countless thousands every year. The guacharo spends its days a half mile or more deep in the interior of mountain caves. Here it roosts and builds its nests in crevices high in the rock walls. It leaves in groups of twenty to thirty shortly after dusk and apparently spends the whole night foraging for food, sometimes covering as much as 200 miles.

Like the cave bat, it seems to have no difficulty finding its way in absolute darkness. An explanation of this ability, acoustic orientation, has been reported by Dr. Donald R. Griffin of Cornell University. The birds apparently are guided by echos of specific sharp "clicking" sounds which they make.

"The individual click," Dr. Griffin explains, "consists of a very few sound waves having a frequency of about 7,000 cycles per second. The duration of each click is about a millisecond (1,000th of a second). The clicks were loud enough to be audible easily about 200 yards inside the cave. Except for their lower frequency, these sounds are very similar to those used by insectivorous bats for their acoustic orientation.

"The external ear canals of three captive birds were plugged with cotton. They then became disoriented when flying in the dark. They collided with every object they encountered. Before and immediately after this treatment they flew about in a small dark room avoiding all collisions with the walls."

Their best known habitat is the guacharo cave in Venezuela's Humboldt National Park, where they are rigidly protected. Most of them nest in a vast subterranean hall more than a half mile long and a hundred feet high. Here more than a thousand of the birds greet the intruder instantly with a wave of awesome and deafening shrieks.

"With the advent of dusk," reports Dr. Eugenio de Bellard Pietri—Venezuelan cave explorer, "the birds come out in compact groups but before the exodus a preliminary flight is held by a few as if to make sure that night is falling. Soon they return to the depths of their somber mansion, evidently to give the flock the all clear signal. Late in the evening there is not a single adult specimen left in the cave. The flight of these birds is silent and cannot easily be detected."

WHERE SNAILS BECOME FLOWERS

The lowly snail reaches an apotheosis—rivalling flowers and butterflies as an expression of nature's artistry—in Cuban forests. Delicate sunrise tints of pink, blue, violet, green and yellow make the shells of two or three genera of tree-dwelling mollusks like rare jewels. Most conspicuous are snails of the genus Polymita, confined to the Oriente province. Here they cover some trees so completely that the effect is like that of a tree of flowers. Only upon close observation can one detect that the blossoms are shells.

The animals live for the most part on a fungus that grows on the bark. The colors of the shells are affected by various chemical constituents of the bark, notably tannic acid, and serve as warning to other creatures. In taste the snails are very bitter and no bird will intentionally attack them. The color serves notice that only a disgusting mouthful is to be had.

Two of the most beautiful of these shell forms were recently discovered by Dr. Paul Bartsch, former Smithsonian curator of mollusks. Fragile, translucent, colored as delicately as the loveliest of orchids, these particular snails are the fairies of the mollusk world in the unconscious artistry with which they have constructed their moving palaces. One, a hitherto unknown species, has a remarkable combination of pale orange, orange buff, deeper orange and flame color—all shading delicately into each other.

The color effect is such as one might find rarely in rose petals. Another has a blending of ivory, olive green, lemon yellow and orange.

TERMITES THAT EAT LEAD

On Barro Colorado island in the Panama Canal Zone the Smithsonian Institution maintains an "experimental cemetery." It consists of rows of upright posts which look like gravestones, half buried in the soil. The purpose is to test the propensities of the island's 42 species of termites—just about man's most persistent and expensive enemy in the tropics—to eat different kinds of wood impregnated with different kinds of repellants and poisons. To date approximately 35,000 tests have been made. The longer the work is continued the more Dr. James Zetek, former director of the station, is impressed with the contrariness and ingenuity of the blind, ant-like insects which achieve sub-human acmes of engineering ability, and whose appetites are marvelous.

Among Barro Colorado's termites are some extraordinary bugs indeed. One, for example, eats lead. It gnaws its way through the lead sheathings on cables. This is not because it likes a lead diet. Lead, in fact, is indigestible and the insects starve to death. But their appetites are so insatiable that the little creatures just keep on gnawing, in the hope that there will be wood on the other side.

This particular insect is known by the scientific name of coptotermes niger. It has been known to eat through a concrete floor nearly five inches thick—again not because of any particular liking for concrete but because of the expectation of coming eventually to digestible wood. The feat was made possible because the sand used in making the concrete contained many fragments of sea shells which were dissolved by a powerful chemical excreted by the insects.

It is very difficult to dispose of termites by poison—that is, permanently. Races have risen here, for example, which seem to thrive on arsenic. The insect lives on the cellulose in wood. This must be digested by certain intestinal bacteria in the digestive tract. If these microörganisms can be poisoned the termite starves. At first at least 99 percent of the bacteria succumb to heavy doses of arsenic. This means that 99 percent of the termites are killed. But always there are a few exceptionally tough bacteria with a high resistance to the poison. Their descendants in a few generations apparently become almost entirely resistant. With their help a new race of termites comes into existence.

Ordinarily termites attack only dead or dying wood. Some of them, however, carry fungi around with them to kill their own wood. The Canal Zone insects can dispose of living trees. Dr. Zetek tells of one attempt to establish an avocado plantation. He warned against it. When the trees had reached the fruit-bearing stage and seemed healthy he was ridiculed for his warnings. Branches were heavy with avocados and there was promise of a record crop. He shook his head when shown the flourishing orchard. "The poor trees," he remarked. "They know they are going to die. They are just making one last mighty effort to preserve their species by producing plenty of fruit and seeds." He secured the orchard owner's permission to chop down one tree. The whole inside, he found, was riddled with termite galleries. This tree and all the others in the orchard were dead within a year.

THE PLANT THAT EATS ANIMALS

There are life-and-death battles in the microscopic world between tiny shelled animals and flesh-devouring fungi. The phenomenon can be compared to that of a tree catching and eating big turtles.

When a culture of diseased plant roots is made, there soon appear great numbers of microscopic plants and animals—bacteria, fungi, amoebae, nematodes and other life forms. Immediately the struggle for survival starts. The animals try to eat the plants and the plants attempt to devour the animals.

Among the animal forms which appear are vast numbers of creatures known as rhizopods. Practically unknown except to specialists, these microscopic creatures play an important part in the economy of life. They are probably the best-equipped of all the new arrivals to survive, since their soft bodies are covered with relatively heavy shells.

Some years ago Dr. Charles Dreschler of the U.S. Department of Agriculture reported the existence of predaceous meat-eating fungi—parasitic forms of plant life—which literally lassoed such unprotected animals as amoebae and thread-like nematodes and proceeded to devour them at leisure by the process of infiltrating their bodies. It would appear that the armored rhizopods are completely protected from these ferocious plants.

But the animal has one weak spot in its defense. It must get its mouth outside its shell in order to eat. Apparently the most inviting forage at hand is the innocent-appearing fungus. The rhizopod proceeds to suck at it with movements which Dr. Dreschler describes as similar to "sucking an egg."

The rhizopod mouth is small. Once it has sucked in any of the fungus its fate is sealed, for, explains Dr. Dreschler, "to such undiscriminating voracity the fungus responds by rapidly proliferating from the partly ingested portion a bulbous outgrowth slightly larger than the mouth, so that the rhizopod is held securely."

The unfortunate shelled animal is like a fish caught on a hook. It struggles vainly to get away. It rushes, but the fungus simply lets out the line until the rhizopod is brought to an abrupt stop and can be hauled in. The line is a filament connecting the body of the fungus with the bulb in the animal's mouth.

Once its prey is secure, the fungus proceeds to send out growths from the bulb through the creature's flesh, literally eating it alive. Very rarely, like a hooked fish, a rhizopod is able to break away.

In the course of its life, a single one of these thread-like fungi will capture many of the shelled animals, lining them up securely mouth-to-mouth on both sides of itself. It absorbs their substance at its leisure. Other predaceous fungi have definite external organs for capturing their prey. This particular species, however, has no external appendages and appears completely inert and innocent until it is stimulated to action by the sucking of the rhizopod.

THE OCEAN'S SOUND BARRIER

A densely woven carpet of life covers the floor of the world of light under the sea—just below the level reached by the most penetrating rays of the sun. It is a carpet of many colors and of flashing lights, the strands of its texture rapidly moving, predaceous, warring organisms. They probably are a mixture of lantern-carrying fish, ten-tentacled squid with malevolent red eyes, and small, luminous, shrimp-like creatures known as euphasids. Their nature can only be deduced by the echoes of sound from their bodies.

This carpet, about 300 feet thick, is the sea's "false bottom." It was discovered by Navy ships making depth soundings during the war. Such soundings depend on the time taken for echoes to be reflected to the surface from the ocean floor. Recorded on a ship's instruments, they represent an extremely precise procedure perfected to the point where a continuous record of depth can be obtained with an accuracy of a few inches.

But, using certain wavelengths of sound, echoes were received from depths between 1,000 and 1,500 feet, whereas the sea itself was known to be two or three miles deep at these places. The only plausible explanation was that there were vast multitudes of floating or swimming objects of some sort, constituting almost a solid surface, at the depths from which the echoes came. The mystery was increased by the fact that the false bottom existed only during daylight. The carpet was laid shortly after sunrise and rolled up at twilight. The indication was that the echo-producing objects rose to the surface at the beginning of darkness—a clue which has given rise to much speculation and argument.

The carpet is under all the oceans, even the nethermost Antarctic. In some areas it seems practically continuous over thousands of square miles. In others it is broken up into smaller areas, like scatter rugs on a floor.

The false bottom is almost as much a mystery today as when it first puzzled the Navy's navigators. All are agreed that it must be composed of vast hordes of animals. They are not directly observable by any known technique. Some indication of their size and abundance, however, can be deduced from the wave lengths of sound which they echo. There must be, it has been calculated, from ten to twenty of these organisms in each cubic meter of water. They echo only long sound waves. High frequency sound passes through them like light through glass and is bounced back from the true sea bottom. They have been a mild nuisance, but never a peril, to modern navigators.

Whatever the organisms may be, they evidently cannot endure any light. At dawn they sink immediately from within about 100 feet of the surface through the zone of moonlight-pale, green illumination which represents sunshine's deepest penetration of sea water.

Chief proponents of the theory that a preponderance of them are squid are oceanographers of the Navy's Hydrographic Office. It is well established that the deep sea abounds in these fantastic mollusks. They rarely are seen at the surface. They move through the water very rapidly by a kind of jet propulsion, gulping water in the mouth and shooting it out explosively from the rear. They are little affected by changes in hydrostatic pressure, as are fish with air bladders. When the false bottom rises at sunset it comes to the surface at a rate of forty to fifty feet a minute. No swimming fish, it is maintained, could rise so rapidly through the decreasing pressure. It would get the "bends", like a human diver brought to the surface in too great a hurry.

These squid range in length from three or four inches to more than a foot. They are of about the right size to return some of the echoes which have been observed. The faintly luminous euphasid shrimps also are known to be very abundant in the depths. Presumably they provide most of the squids' food.

The principal investigations have been carried out by the Navy's Electronics Laboratory and the Scripps Institute of Oceanography of San

Diego. An outstanding difficulty hitherto has been that the echoes have been known only from the false bottom as a whole. They have covered a wide spectrum of sound wavelengths. A recently developed technique is to lower a hydrophone connected with a sound-producing mechanism into the depths in order to record echoes from individual objects at distances of a few feet. Indications to date are that some of them are from a foot to eighteen inches long—too large to be squid and far too large to be shrimp. They can only, it is deduced, be deep water fish. If a great number of fairly large fish are indicated, this false bottom might turn out to be the richest pasture in the ocean for the production of food for man.

Navy divers have swum through the false bottom at night when it was within less than 200 feet of the surface. They have observed enormous numbers of euphasids and other small organisms—but very few fish. This, however, is only suggestive. There is no good reason to believe the carpet has the same texture at night as by day. It is quite likely that the organisms disperse widely over the surface waters.

SNAKES THAT ACT AND LOOK LIKE WORMS

There are snakes that look like snarls of six-inch-long pieces of wrapping twine. These worm snakes are the world's closest imitators of worms. Among the most secretive of living things, they rarely come in contact with man. When they are seen they usually are mistaken for worms. Only zoologists can put them in their true families. These living strings live exclusively under the earth, sometimes in tangled snarls of scores of individuals.

They are the smallest of snakes. Their closest relatives, however, are the gigantic boas and pythons. Judging from their wide distribution—on such isolated spots, for example, as Christmas Island in the Indian Ocean—they are quite ancient reptiles whose wanderings started about fifty million years ago.

They are found most often in termite nests, where they eat the eggs and possibly the larvae. Small earthworms and other soil creatures add to their diet. The worm snakes are almost toothless. Eyes are buried under skin, are only faint spots, and probably only can discriminate light from darkness. The tail looks somewhat like the head—a likeness presumably developed as a camouflage. They retain a snake's scales, but these are highly polished so they can be of no help in crawling.

These Typhlopidae and Glauconidae, as the two major groups are known, are extremely active. When they are exhumed they start at once to burrow back and have been found as much as two feet underground.

Occasionally they may be found in mole holes or in rotten wood where they feed on insect larvae and also, it is likely, get some warmth from the decay process. The snout is used in burrowing. They are hard to hold in the hand, owing to the high polish of the scales. There are approximately 100 species scattered over the world, two coming as far north as the Texas border. They have teeth in only one jaw—the upper jaw for Typhlopidae, the lower for Glauconidae.

A PORCUPINE OF THE SEA

Among the weirdest creatures of the deep is also one of the latest to become known to science—the sea urchin (closely related to star fish) astropyga magnifica. It is the largest sea urchin yet found in the Atlantic. It has approximately 200 bright blue eyes arranged in double rows. The body is covered with several hundred sharp, barbed black spines nearly a foot long.

That so conspicuous an animal, living in such a densely populated region—one of the most intensively studied in the world by biologists—should have remained undiscovered so long probably is due to two reasons. First, if its habits are at all comparable to those of its nearest relatives, it is strictly nocturnal and comes out to forage on the coral sands of the shallow sea bottom only after light has ceased to penetrate the water. During the day the creatures remain secluded, often congregated in great numbers, in holes and caves of the sea floor and under the coral.

Second, it is quite similar in appearance to another smaller member of the sea urchin race with spines as much as 18 inches long which is greatly dreaded and is even reputed to have caused the death of children who have fallen on it. Anybody coming upon a daytime bed-chamber of these fantastic creatures would be likely to leave them strictly alone.

This particular sea urchin is especially interesting in the development of its eyes. These appear to be true sight organs. If a hand is placed in the water near one of the animals the long barbs immediately are pointed in the direction of the intrusion, and as the hand moves the barbs move. Such a creature is practically impregnable. It never, however, takes the offensive. It cannot "throw" its barbs, but they enter the flesh easily and cause painful local irritation. Some species inject a virulent poison which may even kill a human being. There is no evidence that this species is toxic.

Astropyga magnifica, which has more the appearance of a porcupine than of any other land animal, is a scavenger of the sea bottom. It gathers and devours the accumulated debris that falls through the water. It never kills its own food, so far as is known. It has five sharp teeth in its mouth, located on its under surface, with which it can chew away the flesh of dead animals.

This sea porcupine has a peculiar system of locomotion in common with most of its relatives. It has literally thousands of sucker-like feet, which are hollow and attached to tubes within its shell. It moves by forcing water through the tubes and into the particular "feet" which it wishes to use. When these are out of use they are contracted by withdrawing the water. Being a radially symmetrical animal, the creature can move with equal ease in any direction. It has no head—that is, the development of its nervous system and the direction of its locomotion are not fixed in a forward direction, as is the case with vertebrates and insects.

Some members of the sea urchin family have hoof-like formations on the ends of some of their spines, with which they are enabled to walk over the sea bottom without using the suction disks. About the only enemy of these fearsome nightmares of the deep is man. Some species are used extensively for human food, notably among the Mediterranean coast and in the West Indies. The developing eggs are taken from the body and eaten either raw or cooked. Even if it should prove suitable for human food, it is unlikely that the sea porcupine ever will be a rival in this respect of its rival, the "sea rabbit." It is too secluded in its habitat.

WORMS THAT ARE UNKILLABLE

In nematodes life may have reached its greatest capacity for survival. The remarkable persistence of these soil worms has been studied by C. W. McBeth, researcher of the Shell Oil Company. One form, he reports, has been known to survive after 25 years in a glass bottle in a laboratory. Another, a pest of wheat kernels, apparently came back to life after 28 years in laboratory storage. A nematode which had invaded a rye plant, collected in Kansas in 1906, revived after 39 years of complete dehydration in a herbarium.

Those which live as active feeders in the soil, however, are not particularly long-lived. Each species depends on a certain plant type and must starve if this is not available. The recently introduced golden nematode of potatoes, a particularly obnoxious pest, is known, however, to survive as much as ten years in soils where no potatoes are planted. A great mass of eggs is produced, but not laid. They are retained in the body of the mother, who dies. Her skin remains—a bag filled with eggs.

This stays in the soil, apparently unharmed by changing conditions, until potatoes are planted again. Then some mysterious influence, as yet unexplained, causes the eggs to hatch and the whole nematode cycle begins once more.

Due to such a strange tenacity of life this nematode is about the hardest of pests to control. It refuses to stay dead. Other species likewise are specialized in one or more ways of survival under adverse conditions.

Because of the complexity and minuteness of the nematodes, it has been very difficult to determine the effects of heat, cold, flooding and drying on different species. These vary for each. One nematode species, especially resistant to drying, has a skin consisting of nine layers. The ability of this skin to hold moisture inside the minute body undoubtedly is an important defense mechanism. Some species are entirely marine, others are parasites within the bodies of other animals. It has been found that both of these varieties possess skins which are much more permeable to moisture. The original home of the phylum probably was in the sea, but a moisture-proof cuticle has been developed by those which have invaded the land.

The whole body structure of the plant nematode is almost ideally suited to life in the soil. The typical eel-shaped body is well-adapted for moving in the moisture surrounding soil particles. Deviations from this eel-form in certain stages of some species, usually in mature females, are found only in sedentary stages. The larvae and males retain the ancestral shapes. Another deviation is found in the so-called "ring nematodes" which have short, plump bodies incapable of locomotion in the typical whip-like fashion. Movement is accomplished by alternate expansion and contraction of the body.

A majority of nematodes spend a greater part of their lives in the soil. A few, however, are carried from plant to plant by insects. Although moisture is necessary if the tiny animals are to remain active, the soil seldom becomes too dry for them except in the top two or three inches. Their structure is well-adapted for moving up and down.

THE REMARKABLE BRACHIOPODS

A part of the fantastic living world of 200,000,000 years ago has been dissolved out of about thirty tons of yellowish-brown limestone by a Smithsonian paleontologist.

The rock comes from a low mountain range in southwestern Texas—the Glass Mountains, about 250 miles east of El Paso. During the Permean geological period, when some of the earliest known forms of animal life appeared on land, the site of the Glass Mountains was a muddy bottom, probably close to the shore of a warm sea. A bewildering array of animals lived in that sea. They died and eventually were buried in the mud. In some cases their bodies were covered with silica. In others silica replaced the shells. When these rocks are placed in hydrochloric acid the limestone is eaten away but the silica shells remain. Years of skilled labor would be required to chip out of the rock what is obtained in a few days in the acid bath.

Most abundant animals of the ancient Texas sea were the brachiopods or lampshells—essentially shelled worms. The broad road of life is strewn with derelicts, stragglers and deserters. Among the most notable among them are these obscure creatures which, in numbers and apparent prosperity, seem to have been close to the dominant animals in the world in the days when giant amphibians, remotely related to present frogs and toads, and monster scorpions were establishing themselves on dry land.

Brachiopods were among the first animals to leave any traces on earth a half billion years ago. Even at that time they were complex creatures, with nerves and stomachs, which indicate a long ancestry before they left any fossil remains. In the tepid Permian seas they reached their climax in numbers and variety. They survive today, but only in a few places. For all practical purposes they are now among the most obscure animals in existence. In the whole world there are about 110 extant species compared to nearly 500 which Dr. G. Arthur Cooper, Smithsonian Institution curator of invertebrate paleontology, and his associates have obtained from one small area of the Glass Mountain limestones.

The existing brachiopod might be mistaken for a small clam. Zoologically, it is an intermediate form between mollusks and annelid worms, and somewhat closer to the latter than the former. Its way of life actually is nearer to that of an oyster than to that of most worms. It now is believed to be most closely related, through some unknown common ancestor, to the bryozoa or lace weavers. In the past both were classified together. The brachiopod never has become a colonial animal.

Its body is enclosed completely in a shell, secreted by the skin or "mantle", except for a muscular, stalk-like extension, the peduncle, by which it attaches itself to the sea bottom. Inside the shell, folded around the mouth when the animal is at rest, are two arms or tentacles with which it can probe the water and obtain minute food particles. It also apparently breathes through these tentacles, which have a rapid blood circulation.

Most numerous of the extant brachiopods is a curious animal, the lingula, which is nearly world-wide in distribution and whose peduncle is used for food in both Japan and the Philippine Islands. Along the Atlantic coast it is present from Chesapeake Bay to Florida. It makes a nearly vertical burrow in mud or sand from two to twelve inches deep—within which it lives, attached to the bottom by the peduncle. On this footlike appendage it can lift itself until the front part of the shell-enclosed portion of the body is above the surface. This is withdrawn into the burrow instantly on the slightest alarm. The animal apparently has a quite sensitive, although very primitive, nervous system.

The extant brachiopods are usually small animals but in their Permian heyday some attained a length of more than six inches. For essentially 200,000,000 years they were without much competition in the mud burrows to which they had resorted. During this time arose clams, sea snails, and other mollusks which were free to move about and competed with them for the available food supply. The brachiopod was unable to meet this vigorous competition and in a few million years the race was well on its way towards extinction. Most species disappeared. A few, including the Lingula, survived into the age of the great dinosaurs, and their descendants constitute the species living today. They are now obscure creatures and a poverty-stricken group compared to their ancestors.

In the Permian seas they had surplus energy to expend not only in variation of form and habit—but in shell artistry. Some of the specimens obtained by the Smithsonian paleontologists are like glittering gems surrounded by silvery, hair-like spines.

These spiny brachiopods constitute about two-thirds of all the fossils obtained from the Glass Mountain rocks. Although the most abundant they were far from the dominant animals of the Permian sea. They always were defenseless little creatures, dependent on their hard, spiny shells for protection. The sea monsters of the day, creatures related to the present chambered nautilus and some of which were nearly two feet in diameter, unquestionably were the lords of this marine creation. But they were free-swimming predators who had little reason for concern with the humble mud-dwellers. Next to the brachiopods in numbers and variety, and probably their chief competitors, were the ancient lace weavers. Both shared forests of sponges which grew like small trees, up to heights of four feet and four to six inches across. Clams, some of which reached the size of giants, were beginning to claim dominion of the offshore mud and the brachiopods were near the end of their prosperous days.

Like the sedentary worms, and most of the mollusks the brachiopod starts life as a minute, free-swimming, wormlike larva, top-shaped and extremely active. During this period the mortality of the tiny unprotected creatures is very great, but once the mud-dwelling phase of existence has started, the race is secure from most enemies.

FEATHERS ON BIRDS ADAPT TO THE SEASONS

There is a definite seasonal variation in the number of feathers on most birds. It amounts to a "natural adjustment in dress to the needs of the season". This fact has been determined through the laborious process of actually counting the feathers of birds of the same species at different seasons.

The number of feathers declines steadily from early spring until the end of summer when the so-called "post-nuptial" moult takes place, after which the bird gets a new coat to last it a year. The bulk of the new feathers are acquired at the same time, but some are added progressively as the weather gets colder. An exception to this is found, however, among those birds which migrate south early. These apparently get a complete new outfit for their journey, since they will not be obliged to experience any noteworthy change of climate.

WHY THE DODO BECAME EXTINCT

Smithsonian ornithologists have "rebuilt" a dodo. The dodo was a large, pigeon-like, flightless bird which was abundant on Mauritius and neighboring islands in the Indian ocean during the seventeenth century. It became a symbol—first of stupidity and later of extinction.

In its restricted environment it apparently had known no serious enemies prior to the coming of man. It had grown heavy, taken to a ground existence, and lost the ability to fly. It showed no fear of man and, because of its clumsy movements, was easy to catch and slaughter, but its flesh was tough and tasteless, even for sailors who had gone for months without fresh meat. Dutch navigators called it "the nauseating fowl".

Dogs brought by the sailors killed great numbers of the stupid birds. They might have survived despite their slowness and stupidity, however, had it not been for the pigs and Ceylonese monkeys which came to Mauritius with the first settlers. The rooting swine destroyed the bird's eggs and the monkeys devoured its young. It was entirely extinct at the start of the eighteenth century.

THE SHARK OF THE SOIL

There is a protozoan, wormlike monster of the microscopic world, seen only about forty times in two centuries, which gobbles up its fellow one-celled creatures a hundred at a time, walks backwards and forwards at once, and hunts in packs.

It is fifty times the size of the most familiar of one-celled animals, the paramecia, which constitute the dominant population (in numbers) of the invisible creation. It moves among the paramecia like a giant, flesh-eating dinosaur among humans. It is a cumbersome, slow-moving mass of protoplasm. Two or three get together and completely surround a large school of paramecia and these are divided as meals for the captors.

The creature was first described by the Swedish botanist Linnaeus in 1775. He called it *Chaos chaos*. It consists of a single cell, but differs from other one-celled animals in having three cell nuclei, instead of a single one. To reproduce, it splits in three parts, each a new animal.

Chaos chaos moves by stretching itself out into a ribbon-like form and proceeds, by a series of tugs of war, with one end or the other winning out. The animal supposedly is very rare and has been seen only about once every ten years. It may be a missing link between single- and multi-celled animals—or it may be on an entirely different evolutionary track.

THE SLEEPING HABITS OF MAMMALS

The tiny elephant shrew (its elongated nose gives it the appearance of a miniature elephant) apparently never closes its eyes. It is a desert animal, continually exposed to danger, and must "see" even when it is asleep.

Soundest sleepers are the burrowing animals, even when they take their naps above ground. They are conditioned through innumerable generations of safe slumber in their subterranean chambers. Sleeping pocket mice and hamsters can be picked up without being awakened.

Sleep habits appear to be well adjusted to the needs of each species. Most bats, for example, sleep hanging head downward, suspended by the nails of the hind feet. This places them in a good position for sudden flight at any alarm. They have only to let go with their toes and spread their wings.

Curious sleepers are the armadillos. They tremble almost continually in their sleep.

THE EERIE EYES OF ANIMALS AT NIGHT

Eerie lights shine in the silent blackness of the jungle night. There are red lights and green lights, orange lights and yellow lights. They are reflections from the eyes of all sorts of animals.

This weird phenomenon has been observed closely for some years by Ernest P. Walker of the National Zoo in Washington. The shining of eyes is a fairly well-known phenomenon but most of the observations have been made in the wild. The owner of the eyes is usually unknown, and it is virtually impossible to observe the animal again. Mr. Walker has concentrated his observations on caged animals.

He uses a reflecting headlamp, similar to a hand flashlight, worn on the forehead and connected with a three-cell battery in his pocket or attached to his belt. This is necessary because the rays of reflected light must parallel closely the line of sight of the observer.

The "shines" range in color from pale silvery through silver, blue-green, pale gold, gold, reddish gold, brown, and amber to pink, with a range of intensity from dull to very brilliant. The eyes of alligators and crocodiles "give one the impression that he is looking into a brilliantly glowing pinkish opening in a dull-surfaced bed of coal". Most eye shines of mammals have the appearance of coming from highly polished metal surfaces.

"Sometimes," explains Mr. Walker, "it is like looking into an incandescent globe of the color indicated. Often pronounced light rays seem to emanate from the eyes. With some eyes, such as those of the smaller rodents, the effect is that of looking into an illuminated piece of amber.

"In the case of animals that have eyes that glow, it appears that we look into the eye through the pupil as if the reflection came from the front surface of the retina. In those animals that give a reflection as if from polished metal I gain no impression of looking into the eye. In most cases the reflection is not obtainable closer than from eight to twenty feet—a distance which prevents one from observing which surface reflects. The reflection from alligators and crocodiles can be seen when the observer is within a foot of the animal."

Most animals stare at light, or barely move their heads. There seldom is any "startle" response when a beam is flashed upon them. There is no shine in the eyes of higher apes and monkeys. There have been reports of something of the sort from human eyes, but no definite proof has been offered. There was a faint suggestion of a reflection from the ring-tailed lemur, a close relative of the monkey family. On the other hand, the most brilliant eye-shine of all was from two tiny members of the lemur tribe, the slow loris and the potto.

The majority of rodent eyes shine dully in browns, hazel or amber. Porcupines are an exception. Their eyes are very brilliant, generally silver and reflecting over a wide angle. Whether snakes have any true eye reflection is questionable. Light is reflected, however, from the surface of the scales over the eyes.

WORLD OF THE BLIND

There is a fifth realm of life—the wet, heavy, black darkness of limestone caves whose chambers, ponds and streams harbor almost a hundred species of worms, pseudo-worms, fish, insects and salamanders which have become adapted to life in this cheerless world over millions of generations.

Nearly all are white and blind. Blind white fish chase and eat blind white worms. Blind white spiders spin nets to trap blind, white flies. All are sluggish creatures. Kentucky's Mammoth Cave alone contains approximately 50 species. Latest to be classified scientifically are small, rather gruesome white worms of the sort one might imagine feeding on the dead. They live in water, clinging to the bottoms of rocks.

Most spectacular of cave animals is the spectral Proteus, found in limestone caves of Dalmatia, Carinthia and Carnolia in southeastern Europe. It is a kind of salamander, related to frogs and toads. It looks and acts like a big white worm. The creature is about a foot long and pure white except for its gills, which are vivid red. There are three pairs of these gills, which look like coarse feathers, just behind the head.

The Proteus spends its whole life in total darkness, and at an almost constant temperature of 50 F. The body is slender and decidedly wormlike, but there are two pairs of very feeble, inconspicuous little legs, placed quite far apart.

Nature has made the Proteus a true creature of darkness—perhaps more so than any land-dwelling worm. As described by the late Dr. Austin H. Clark, Smithsonian Institution biologist: "The Proteus is almost as sensitive to light as a photographic plate. The light of a candle at some distance is strong enough to make it restless. If it is kept in a place from which light is not entirely excluded its white skin turns cloudy with the appearance of gray patches, and if it is kept in an ordinary lighted room it eventually turns jet-black."

Proteus is eyeless. It seems feeble and helpless. Yet it is well adapted for its life in dark caves. Most of the time it lies at the bottoms of pools, completely motionless. But, says Dr. Clark, "any small living thing in the water attracts its immediate attention. It advances toward it, snaps it up and eats it. It seems to be guided mostly by the movements of its victims in the water, possibly also by a sense of smell. In the deep caves food naturally is scarce and the animal often must go for a considerable time without anything to eat. In captivity individuals have lived for months with no food at all."

Ghostly dweller in the everlasting darkness of limestone caves in the Ozarks is the Typhlotrition, a blind, wormlike white salamander of the same general family as Proteus. It is a long, slender, nearly transparent creature, which has evolved a long way towards complete blindness. The newly hatched young have functioning eyes but these degenerate in the adult so that it does not seem able to discriminate light from darkness. It is barely able to stand on its thin, barely visible legs. It lives on blind crustaceans and apparently spends most of its life crawling through the small, underground streams which seep through the limestone rocks of the Ozark foothills.

A quite similar creature of the same family was discovered in 1896 in Texas during the boring of an artesian well. A subterranean stream was struck at a depth of about 200 feet. From it this white, wormlike creature was shot out, together with some remarkable crab-like animals. A single specimen of a similar animal since has been found in Georgia. Both these organisms are more wormlike even than Proteus. They apparently have lived for milleniums in streams flowing hundreds of feet below the earth. Both, it has been conjectured, are larval forms of a well-known salamander of surface waters, which have become permanent larvae. They have lost the ability to undergo metamorphosis, like the change of a tadpole into a frog or a caterpillar into a butterfly.

Most numerous of American limestone cavern animals are white, blind grasshoppers—the cave crickets. They are small insects with antennae about an inch long. With these they feel their way over the dank walls upon which they swarm. Best known are three species of cave fish, minnow-like and from two to three inches long. They have not lost their eyes entirely, although these long since have been sightless. They have compensated for the loss of sight by an extremely acute sense of touch. The slightest movement of the water will send a school of them scurrying for shelter among the rocks. The blind white worms are supposedly their chief food.

None of the cave animals are very aggressive. Their chief nutriment is believed to be organic matter carried by water, which seeps into the dank chambers from the world above, but how they make use of this is unknown. All are quite primitive types which have remained very conservative after their first migration from the world of light into the world of darkness. They are old both racially and in their behavior as individuals. Secure in the black depths, some of them are quite likely to be the last living creatures on earth.

THE REMARKABLE CLAM WORMS

Fantastic giant of the nemertinean race is Cerebratulus lactus, commonly known as "the clam worm" along the Atlantic Coast from Florida to Massachusetts. It is from ten to twelve feet long, can contract to two feet, and is an inch wide. Its favorite dwelling is a burrow six to eight inches below the surface, usually in an old mussel bed among broken shells and stones where it is almost impossible to sink a clam hoe.

Outside the burrows it is seldom seen except occasionally at high tide, gliding among sea weeds or in the shade of rocks in tidal pools. It is unlikely that any burrow is occupied very long, as the nemertinea is moving about constantly through mud in search of food. The animal is highly specialized for burrowing. Ordinarily its "head", or front end, is broad and rounded. By a muscular contraction, however the shape of the head can be made pointed and is thrust forward in the mud, when its normal contour is resumed. Then again comes the muscular contraction, the pointed head, and another thrust forward. This occurs over and over again. The contraction waves follow each other so quickly that the drilling process appears constant. The proboscis does not seem to be used in the actual drilling operation, but is kept probing for points of least resistance and turns aside at the slightest obstacle.

The favorite food of cerebratulus lactus is said to be another abundant burrowing worm, the nereid, which is nearly as large in diameter, belongs to a higher order, and has powerful biting jaws. The victim always is swallowed tail first. Its burrow is a U-shaped tube in which it is unable to turn around. The nemertean probes through the mud for the tail end in such a burrow. The nereid, seized from behind, cannot bring its fighting apparatus into use. Actually, however, it never appears to struggle against being swallowed—a remarkable fact since nereids fight fiercely among themselves. The reason, it has been postulated, is that the victim's nervous system is paralyzed by the poisonous slime excreted by cerebratulus. When a minute drop of this is placed on the tongue, it parches the whole mouth and the intensely bitter taste remains a long time. The worm requires about ten minutes to swallow a nereid, but by that time the prey is half-digested. The flow of this mucous is quite copious. When several healthy worms are placed in a pail, the bottom is soon filled with a hardening mass of it from which the animals must be cut or pulled. When crawling, the worm exudes a mucous trail, like a snail.

A comparable Mediterranean species, Nemertes borlasi, was described by the French naturalist Quatrefages:

"This gigantic worm is from thirty to forty feet long, brown or violet, and shining as varnished leather. It lurks under stones and in hollows of rocks where it may be met with, rolled into a ball and coiled in a thousand seemingly inextricable knots which it is incessantly loosening and tightening by contraction of its muscles. The animal is nourished by sucking a kind of small oyster which attaches itself to various substances under water. When it has exhausted the food around, it extends its long, dark-colored, riband-like body, which is terminated by a head bearing some likeness to the head of a serpent. It pauses gently, moves from side to side as if endeavoring to investigate the ground, and finally succeeds in finding a stone to suit its purposes about fifteen to twenty feet from its former retreat. It then begins to unwind its coil and arrange itself in a new domicile. In proportion as one knot is loosened, another forms at the opposite extremity."

A report of the Gatty Marine Laboratory of St. Andrews University in Scotland tells of the species Cerebratulus angulatus, which was mistaken for a fish. "But when the fisherman stretched out his hand net to capture it, instantly to his astonishment it shot out to more than a yard long. In the laboratory it swam with undulatory up-and-down movements, as an eel swims laterally."

The nemertinea are a progressive race. Some have invaded the deep sea and some the dry land. They have been obtained from depths of more than 6,000 feet. The deep-sea species have undergone peculiar adaptations for a life of swimming slowly or floating idly at whatever depths they have chosen for their habitat. They have lost their eyes and their brains are quite rudimentary compared with those of their land or shallow-water relatives. All have increased greatly the amount of gelatinous tissue between the internal organs, so that they have a low specific gravity. The deep-sea forms thus far collected are broad and flat. Some have taken on the appearance of small fish with outgrowths on the sides of the body which resemble fins, and with the rear end flattened like a fish's tail. Some have developed tentacles around their mouths.

Most of the ribbon worms of the open sea are nearly transparent. Some, however, are among the most brilliantly colored of the nemertinea race, with coat patterns of yellow, orange, red, and scarlet. Most of these creatures are small, measuring only a fraction of an inch in length. The largest is about six inches long—thus, as one biologist points out, comparing to the smallest like an ox to a mouse. These pelagic species are found in all the oceans. They are carried around the world by deep-sea currents.

About twelve species have abandoned the shore for dry land where they lead active lives and seem to have become almost independent of water.

They cannot, however, endure being completely dried out. They do not make their own burrows, but in periods of drought, it is believed, they make use of earthworm burrows. Some have been found under the dead, damp bark of tropical trees. Their chief food consists of earthworms.

WINGED REPTILE

The largest flying animal the world has known was a winged reptile, the pterodactyl, of a hundred million years ago. It had a wing spread of more than twenty feet, supporting in the air a body which would hardly have weighed more than thirty pounds. Its head was nearly four feet long with a dagger-like, narrow, pointed toothless beak. It lived around the ancient sea which once extended northwestward from the present Gulf of Mexico through most of Kansas. Presumably it lived entirely on fish and made long, gliding flights over the water.

The structure of this reptile, insofar as it could be realized from fragmentary fossil bones, was studied carefully by Dr. Samuel P. Langley while he was at work on early models of his airplane. Did the pterodactyl, Dr. Langley asked in a somewhat pessimistic progress report, represent the best Nature could do in the way of flight? Could man hope to do better than Nature?

VICIOUS FIRE ANTS

One of the most vicious of insects is the fire ant of South America—a small red ant whose sting burns like the point of a red hot pin pushed into the skin. Hordes of these creatures have forced the populace to abandon Brazilian towns. The soil of a village can be completely undermined by the ants. The ground is thoroughly perforated by the entrances to their subterranean galleries.

"The houses are overrun by them," says Edward Bates in *A Naturalist on the Amazon*. "They dispute every fragment of food with the inhabitants and destroy clothing for the sake of the starch. All eatables must be suspended from rafters in baskets, with the cords well soaked in balsam, the only known means of preventing the ants from climbing. They seem to attack persons out of sheer malice. If we stood for a few hours in the street, even at a distance from their nests, we were sure to be overrun and severely punished. The moment an ant touched the flesh he secured himself with his jaws, doubled his tail, and stung with all his might."

THE ARCHITECTURAL GENIUS OF BIRDS

Birds rival ants and termites as architects. One species builds nests as big as small human dwellings—as much as 25 feet long, 15 feet wide and ten feet high. This is the sociable weaver bird of the desert western areas of South Africa. Such an apartment house, woven out of sticks and straw, may contain as many as 95 individual nests. It is the community product of a flock of from 75 to 100 pairs. The sheer bulk of the nesting material gathered is striking evidence of the impelling year-round urge of the building instinct.

This bird, says Dr. Herbert Friedmann, Curator of Birds at the Smithsonian Institution, "is about as sociable as any bird could possibly be. It is always found in flocks, feeds in flocks, and breeds in the large, many-apartmented compound nests. With this extreme sociability and sedentary habit of life the territorial relations of the species have been modified in a way that is quite remarkable, perhaps unique, among birds. Instead of each pair having its own breeding territory, each flock seems to have a definite territory whose boundaries are seldom crossed by individuals of other flocks.

"In an area of approximately 1,000 square miles I found only 26 nests. The flocks ordinarily do not live in very close juxtaposition to each other. The nests are so large, so conspicuous at great distances, and the trees so relatively few in number that I am quite certain I found practically every nest in the area."

In spite of the highly developed communal life, Dr. Friedmann notes, there appears to have been no break-down of the family. Whether each male has one or several mates, however, is unknown. In the construction of the apartments there is some evidence that each family builds its own individual nest, while the whole flock cooperates in constructing a roof over the whole. The structures often become so heavy eventually that they crash to the ground and all the work must be done over.

Woodpeckers that carve "apartment houses" out of hardwood tree trunks have been observed by Dr. Alexander Wetmore in the dark, rain-drenched forests of the La Hotte mountains in Haiti. On one occasion he was astonished to find a dozen pairs going in and out of nests in a single dead tree trunk standing in an open space, the holes being from three to ten meters from the ground and in some cases less than a meter apart. There was no question that the woodpeckers were colonizing, as the trunk was a veritable apartment house with the birds climbing actively over its surface and flying back and forth to the nearby woodland.

In the same mountains Dr. Wetmore found another apartment builder, the palm chit-chat. It is a gregarious species that lives in small bands, each being made up of several pairs having a communal nest as the center of its activities. The largest bands frequenting a single nest do not appear to contain more than 20 birds.

The nests are constructed of twigs about the size of a pencil and from ten to 17 inches in length. The bird itself is only seven or eight inches long. Yet it is able to carry these heavy "timbers" 30 or 40 feet from the ground. One of the nests examined was about the size of a bushel basket and evidently was occupied by only a few pairs. There was a roughly defined central tunnel four to five inches in diameter leading through the mass of sticks and opening to the outside at either end. Near each end was a slight accumulation of bark that made a little platform.

The "apartments" opened from the tunnel on each side. There was a central chamber, supposedly a community room, about five inches in diameter, its floor carpeted with fine shreds of bark. Each nest was a separate unit, with its own door to the outside. There were, however, roughly defined passages running through the interlacing twigs at the top of the nests that permitted the birds to creep about under cover.

One of the most intricate of all bird nests is that of the South African penuline titmouse, distantly related to the American chickadees. It is made of a wool-like plant fiber, very intricately and delicately woven. The form is that of a small bag hanging from a thorn bush. It has one visible opening, a false one which leads nowhere and apparently is intended entirely as camouflage. The real entrance is skillfully hidden, its location known only to the builder. When the mother bird enters the nest she lifts a concealed flap, slips through, and closes it behind her. She again closes it just as carefully when she leaves the nest. There is not the slightest indication on the surface of the finely woven fiber of the existence of the flap.

The Ceylon tailor bird, orthotomus sutorius, makes its nest by actually sewing large leaves together in the shape of a horn, using its bill as a needle. As described by the British naturalist A. G. Pinto: "The first thing she did was to make with her sharp little beak a number of punctures along each edge of the leaf. Having thus prepared the leaf, she disappeared for a little and returned with a strand of cobweb. One end of this she wound around the narrow part of the leaf that separated one of the punctures from the edge. Having done this she carried the loose end of the strand across the under surface of the leaf to a puncture on the opposite side where she attached it to the leaf, and thus drew the two edges a little way together. She then proceeded to connect most of the other punctures with those opposite them, so that the leaf took the form of a tunnel converging to a point. The under surface of the leaf formed the roof and sides of the tunnel. There was no floor to this, since the edges of the leaf did not meet below, the gap between them being bridged by strands of cobweb.

"When lining the nest the bird made a number of punctures in the body of the leaf, through which she poked the lining with her beak, the object being to keep it in situ. All this time the margins of the leaf that formed the nest had been held together by the thinnest strands of cobweb, and it is a mystery how they could have stood the strain. However, before the lining was completed the bird proceeded to strengthen them by connecting the punctures on opposite edges of the leaf with threads of cotton. She would push one end of a thread through a puncture. The cotton used is soft and frays easily so that the part of it forced through a tiny aperture issues as a fluffy knob, which looks like a knot and usually is taken as such. As a matter of fact, the bird makes no knots. She merely forces a portion of the cotton strand through a puncture and the silicon in the leaf catches the strands and prevents them from slipping. Sometimes the cotton threads are long enough to admit of their being passed to and fro, in which case the bird uses the full length."

The leaves are not killed by the tailoring process and remain green. Hence the nest is almost impossible to detect.

THE FEROCIOUS LEECH WORMS

Armies of billions of ferocious worms defended and preserved a fabulous 1,000-year-old Arabian Nights kingdom for three centuries. This kingdom was templed Kandy in the center of Ceylon, encircled by low, densely forested mountains. It was the site of one of the most picturesque ancient civilizations of the Orient which had degenerated into a brutal despotism when the first European invaders, the Portuguese, came to the island early in the sixteenth century.

Armed with arquebuses, the white man established missions and trading posts on the coast with little difficulty, but the forested mountains proved impassable. The Portuguese soldiers were hard put to pitch their camps in deep jungle bush and in bug-filled marshes. Grass and bushes swarmed with little green worms—extremely nimble creatures about an inch long which subsisted on the blood of warm-blooded animals. They seemed to prefer human blood. They attacked the soldiers night and day. Clothes were no protection. The worms dropped in streams of blood from eyelids and ears. They swarmed on all sides in ever-increasing numbers as the invading forces penetrated further into the jungle. With no defense against this unanticipated enemy, the Europeans were forced to retreat long before the temples of Kandy were in sight. They made no further effort to conquer the ancient kingdom.

The Dutchmen who followed the Portuguese were content to remain in their barricaded coastal trading posts. A century later came the British East

India Company with a small army of Sepoys commanded by British officers. The ruler of Kandy, quite secure within his green-worm defenses, was Raja Sinha, one of the cruelest of Oriental despots. He spurned all overtures at negotiation with officers of the trading company.

Once again his kingdom was invaded. During the march into the mountains the Sepoy soldiers suffered so badly from the attacks of the worms that some died and many others deserted. The force was so badly depleted that further advance became impossible. Only when British regulars took over the invasion years later was an armed force of white men able to reach Kandy. Previously only individuals, chiefly Portuguese Franciscans, had been able to cross the terrible green-worm barrier.

Sir Emerson Tennent, British historian of Ceylon, describes these worms as normally about an inch long, slender as needles, and able to stretch their bodies to double the ordinary length. Ceylonese natives had been able to protect themselves to some extent by smearing their bodies with lemon juice and tobacco ashes.

"On descrying the prey," says Tennent, "they advance rapidly by semicircular strides, fixing one end firmly and arching the other forward until by successive advances they can lay hold of the traveller's foot, when they disengage from the ground and ascend his dress in search of an aperture. The wound they make is so skillfully punctured that the first intimation is the trickling of blood or the chill feeling of the worm as it begins to land heavily on the skin."

These worms, hirudinae or leeches, are remotely related to earthworms with a quite similar internal structure, but highly specialized for an exclusive diet of warm blood which they take from any mammal that comes within reach. The blood-sucking species—not all species are this type—have triangular mouths with extremely sharp chitinous [of the same material as the shells of insects] teeth. The bite, so rapidly and skillfully administered that it seldom is felt, has been described as resembling the movement of a circular saw. Haemadipoa, the Ceylon species, described by Tennent, reportedly has five pairs of keen eyes and as many as 100 body segments. All the blood eaters have two suckers, one on the front and one on the rear of the body, by means of which they cling to their victims. All have the ability to contract the body to a plump, pear-like form and extend it to a wormlike form.

The green worms are as much of a terror as ever to travelers in Asian jungles. A species akin to that of the Kandy defense armies guards the thickly forested approaches to the Himalayas in Nepal It is described by Dr. George Moore, chief of the United Nations medical mission to Nepal:

"These leeches, little segmented worms about two inches long, were particularly provoking and troublesome until our team reached an altitude of 14,000 feet. Along the trails, on each ledge leading to the pass, leeches would lie in the shade and moisture until nearby footsteps vibrated their sense organs. Then they would inch from rock to rock at incredible speed, traveling their entire length toward the sound in about a second and then stopping to perch on the rock with their front ends sticking in the air. Immediately they touched a human body they would fasten themselves to it and search for warm skin. Often they would drop from trees. They could penetrate eyelets of shoes and pores of socks by lengthening the entire body. Huge clots of blood would be found on the skin where the greedy worms had fattened themselves to a fragile bursting point."

The leech encountered by Dr. Moore's mission long has been notorious as one of the most vicious animals on earth. It has made some areas of the Himalayan foothills uninhabitable. Travelers and hunters are terrified by it. It exists in incalculable numbers and attacks at least all warm-blooded animals. Horses are driven wild. Cattle and dogs sometimes are blinded and the young and sick killed. It has been known to attack the deadly cobra, striking at the eyes and blinding the reptiles. The respect in which it is held in indicated by its zoological name montivindictus, or "defender of the mountains."

Its stronghold is the highly humid zone at the foot of the Himalayas between altitudes of 4,000 and 6,000 feet. Its period of activity occurs during the rainy season, when it can move freely without danger of drying out. At other times it seldom is seen except at night when grass and bushes are wet with dew.

The worm lurks at the bases of plants. It is stirred to action by the slightest movement of stems or vibration caused by footfalls. An inherent impulse, or geotropism, then impels it to climb any plant or vertical object with which it happens to be in contact. At the top it extends its body horizontally and probes the surroundings.

Once a victim is found, the hungry worm seeks a thin patch of skin richly supplied with blood capillaries. There it attaches itself by means of the cup-like sucker at the front end of its body. Immediately behind this cup are three radiating ridges, or jaws, each provided with about 70 sharp teeth. With these three rows of teeth it cuts three duplicate slits on the skin, meeting at a common center. From the star-shaped wound the warm blood is sucked. Meanwhile from its own glands the leech secretes hirudin, a substance which prevents blood coagulation, and also some as yet unknown substance which preserves blood. The blood is pumped into a storage tank in the leech's stomach. At a single feeding the animal can store up as much as three-fold its own weight. Then it can live as long as three months without another meal.

THE COMPLEX SPIDER'S WEB

A single strand of a spider's web may consist of several thousand separate filaments. On the creature's abdomen are four to six teat-like organs. Each secretes through several hundred extremely minute tubes a viscous fluid which hardens immediately when exposed to air. The spider attaches its abdomen to some solid object and pulls out the threads by moving its body forward. The hind feet are used to bring the hundreds of filaments into a single thread.

MONSTERS OF THE DEEP: THE GREAT SQUIDS

Giants of the mollusk family and about the most loathsomely fantastic creatures on earth are the great squids. One may weigh as much as half a ton. The largest known specimen, a replica of which is among the Smithsonian Institution exhibits, was 55 feet long. It had ten arms, two of them approximately 35 feet long and two-and-a-half inches in diameter. Its eye measured seven by nine inches. Many strange sea serpent stories have been told by persons who merely saw a writhing arm of one of these creatures on the surface. In recent years, however, there has been no reliable report of an encounter with such an animal and it may be close to extinction. Normally it is a denizen of profound depths and darkness and presumably shuns light. It is associated chiefly with the North Atlantic, especially around Newfoundland.

There are not more than a dozen entirely authenticated accounts of seeing the monster. Just after the middle of the last century, Rev. Mr. Harvey of St. Johns, Newfoundland, began to gather "sea devil" reports from fishermen and these constitute a substantial portion of the literature on the subject. He reported that in 1874 two St. Johns fishermen in an open boat observed an object floating in the water which they thought to be wreckage: "When they approached it reared its parrot-like beak, big as a six-gallon keg with which it struck the bottom of the boat violently. It then shot out from around its head two huge, livid arms and began to entwine them around the boat. One of the men seized an axe and cut off both arms as they lay over the gunwale, whereupon the creature moved off and ejected an immense quantity of inky fluid which darkened the water for two or three hundred yards.

"Early in the morning of November 21, 1877," Harvey informed Prof. Addison E. Verrill of Yale, "a big squid was seen on the beach at Trinity

Bay, still alive and struggling desperately to escape. It had been carried in by the tide and a high inshore wind. In its struggles to get off it ploughed a trench or furrow 30 feet long and of considerable depth by the stream of water which it ejected with great force from its syphon. When the tide receded it died. The body was eleven feet long, with tentacle arms 33 feet long. The shorter arms were about eleven feet long."

"In 1878," Harvey reported, "Stephen Sherring, a fisherman residing in Thimble Tickle, was out in a boat with two other men. Not far from shore they observed some bulky object and supposing it might be part of a wreck they moved towards it. To their horror they found themselves close to a huge fish with large, glassy eyes, which was making desperate efforts to escape and churning the water into foam by the motions of its immense arms and tail. It was aground and the tide was ebbing.

"Finding the monster partially disabled, the fishermen plucked up courage and ventured near enough to throw the grapnel of their boat, the sharp flukes of which, having sharp points, sunk into the soft body. To the grapnel they had attached a long rope which they carried ashore and tied to a tree to prevent the fish going out with the tide. His struggles were terrific as he flung his ten arms about in dying agony. Ever and anon the long tentacles darted out like great tongues from the central mass. At length it became exhausted and when the water receded it expired. The body measured twenty feet from the beak to the extremity of the tail. The fishermen, knowing no better, proceeded to convert it to dog meat."

At about the same time H. T. Bennett of English Harbor, Newfoundland, wrote a newspaper account quoted by Prof. Verrill: "A giant cephalopod was run ashore at Coomb's Cove whose body measured ten feet in length and was as big around as a hogshead. One arm 42 feet long and about the size of a man's wrist. The other arms were only six feet long but nine inches in diameter and very stout and strong. The skin and flesh were 2.25 inches thick and reddish inside as well as out. The suction cups were all clustered together near the extremity of the long arm and each cup was surrounded by a serrated edge, almost like the teeth of a handsaw. I presume it made use of this arm for a cable and the cups for anchors when it wanted to come to as well as to secure its prey. This individual, finding a heavy sea was driving it ashore tail first seized hold of a rock and moored itself quite safely until the men pulled it ashore. It was probably a female."

The monstrous ten-tentacled mollusk fights terrible battles with whales and sometimes large parts of tentacles are spewed by leviathan in its death agonies. So far as known only one such battle ever has been witnessed and described. The British author Frank T. Bullen in the *Cruise of the Cachelot* tells of seeing in the South Indian ocean "a very large sperm whale locked in deadly conflict with a cuttlefish almost as large as himself whose interminable tentacles seemed to enlace the whole of his body. The head of the whale seemed a perfect network of writhing arms. It appeared as if the whale had the tail part of the mollusk in his jaws and in a businesslike, methodical way was sawing through it. By the side of the black, columnar head of the whale appeared the head of the great squid, as awful a sight as one could well imagine in a feverish dream. I established it to be as large at least as one of our pipes which contained 350 gallons. The eyes were very remarkable from their size and blackness contrasted with the livid whiteness of the head. They were at least a foot in diameter. All around the combatants were numerous sharks, like jackals round a lion, apparently assisting in the destruction of the huge cephalopod.

"The occasions when these big cuttlefish appear on the surface must be very rare. From their construction they appear fitted only to grope among rocks at the bottom of the ocean. Their normal position is head downward, with tentacles spread like ribs of an umbrella. The two long ones, like the antennae of an insect, rove unceasingly around seeking prey. In the center of the network of living traps is a chasm-like mouth with an enormous parrot-like beak."

"Insatiable nightmares of the sea," the French philosopher Michelet called the creatures. Nothing is known, of course, of their numbers or of their ways of life in the dark depths. The few seen or captured probably have been sick or badly injured. It has been estimated that one female may lay as many 40,000 eggs in a season, but the mortality of eggs and young must be enormous. It is doubtful if one in a million ever becomes a mature animal.

A scarcely less fantastic animal, but more familiar and far less fearsome, is the eight-tentacled octopus. Some of the largest are found off the coast of Alaska. The largest known had arms 16 feet long and a radial spread of 28 feet, but the central body itself was not more than six inches wide and a foot long.

Most familiar of the race is the Mediterranean octopus; its tentacles often are sold for food in Sicilian markets. The largest known was nine feet long and weighed about 50 pounds. This animal reportedly was captured by a fisherman with his bare hands. One specimen found dead on a beach near Nassau had tentacles five feet long and weighed more than 200 pounds.

It is a rather sluggish, timid animal which seeks shelter in holes and crevasses among offshore rocks. It feeds mainly on clams and oysters. When frightened it surrounds itself with a cloud of ink-like fluid. There is no reliable reason to believe it ever attacks man.

THE VANISHING WHIPPOORWILL

Probably not one person in a thousand has ever seen a whippoorwill. Its melancholy song is one of the most familiar chords in the symphony of the summer evening but to the majority of listeners it is only a disembodied voice in the dark. The singer has come about as near to achieving invisibility as any living creature.

The whippoorwill is a migrant bird, spending its winters in Florida and its summers from March to October in the north. It travels entirely at night, sometimes in large flocks. It builds no nest but lays its flecked eggs on the ground depending on the flickering shadows of the woodland over the background of dried leaves to conceal them.

The bird is masterfully camouflaged by nature and usually selects a spot for its eggs where the woodland floor is free of underbrush and the trees are spaced far enough apart to cast an uneven shade. The male presumably sleeps all day while the female sits on the eggs or broods the newly hatched young, but at night he stands guard, may take his turn on the nest, and hunts insects for his mate.

The chick, almost exactly the color of the dead leaves among which it lies, remains essentially invisible. Nests are found only by accident.

Whippoorwills live almost exclusively on night-flying insects, especially moths and mosquitoes. They have been recorded, however, as sometimes hunting for worms, beetles and ants under bark, or on the ground.

The bird makes no particular effort to conceal itself from humans. Apparently it does not regard them as dangerous. There are cases where it actually has lit on the head of a man standing motionless in the dark. The female has been observed to fly about carrying her young between her thighs. She also, it has been reported, sometimes carries them in her bill, but there is no satisfactory evidence of this.

The whippoorwill is fond of taking dust baths. Sometimes one is caught by the lights of an approaching auto as it dusts itself in the middle of a country road.

The bird is remarkable for the regularity of its song and for the number of times the melancholy refrain is repeated without a pause. From 150 to 200 is not unusual. The naturalist John Burroughs claimed once to have counted 1058 such repetitions. The song is continuous from dusk until about 9:30 and from about 2 until dawn. It is heard rarely in the intervening hours.

The whippoorwill, it is pointed out in a Smithsonian report, has come to depend almost exclusively on darkness for its protection. For this reason it has suffered little, as have many other birds, with the cutting away of the forests and the advances of cities. Its enemies in the dark are some hawks, owls and foxes, but has exceptional powers of flight which often enable it to escape even when discovered.

The birds linger in the north only until the first killing frosts which destroy or drive into shelter the insects on which they feed. Then they start their night migrations southward which sometimes carry them as far as Central America.

ANTS CAN SMELL ALMOST ANYTHING

The sense of smell is remarkably acute in all ants—at least equalling that of dogs.

The outstanding ant odor is that of formic acid, which is somewhat like that of illuminating gas, exuded from the bodies of all species. But this is only the smell of the race. It must be subject to an infinite number of variations to most of which ants alone are sensitive. They know their comrades, even after a long separation. Famed naturalist Sir John Lubbock once returned some ants to their old nest after a separation of 21 months. They were amicably received and evidently recognized as friends. On the other hand if a strange ant is placed in a nest of her own species she is at once attacked.

Dr. William M. Wheeler insists that even the human nose can detect some different species and even, in a few cases, different castes by their odors. Thus, over and above the formic acid smell, the smell of one species suggests ether, of another lemon-geranium, and of still another rotten coconuts.

At least one species of ant has three distinct odors: 1. A scent deposited by the feet, forming an individual trail by which she retraces her own steps. 2. An inherent odor of the whole body which is identical for all of the same lineage and a means of recognizing blood relatives. 3. A nest odor, consisting of the commingled odors of all members of the colony, used to distinguish their nest from the nests of aliens.

Evidently the odor of ants changes with age. It has been pointed out that "a cause of feud between ants of the same species living in different communities is a difference of odor arising out of difference of age in the queen whose progeny constitute the communities." Ants apparently not only differentiate the innate odors peculiar to the species, sex, caste and individual, but also the incurred odor of the nest and environment. As worker ants advance in age their progressive odor intensifies or changes to such a degree that they may be said to attain a new odor every two or three months.

FISH THAT FISH FOR FISH

There are fish that fish for fish with worms. That is, they use wormlike appendages of their own bodies, developed through millenia of evolution, to catch worm-eating fellow fishes. This curious quirk of fishing fish is revealed in a bulletin of the International Oceanographic Foundation.

The practice is confined to the pediculati, known as angler fishes. The best known of them lies on the bottom partially concealed in sand or mud. One of the spines of its dorsal fin is extended in the form of a jointed fishing rod. At the end there is a fleshy lump, with a striking resemblance to one of the most tasty marine worms. The fish lies perfectly still with its enormous mouth closed, while the wormlike end of its rod waves to and fro. Other fishes approach the lure until they come within striking range. Then the great mouth opens with remarkable speed and engulfs the prey, which is prevented from escaping by backward-directed teeth.

Some other deep-sea anglers have luminous lures at the tip of the rod, somewhat like a small, light-emitting fish. In the total darkness of deep waters this is fatally attractive. Because of the huge size of the angler's mouth the prey may be almost as large as the fisherman. Other deep-sea fishes dispense with the rod but have light-emitting organs on the sides of the body. These must play some part in attracting other sea animals. Some of these luminous fishes are able to swallow other fishes many times their own size because of their ability to distend their mouths and throats.

About all the ways man has devised for catching fish have been devised by fishes themselves long before man came on the scene. Traps—for example. There is a fish in Florida waters known as the greater sand eel. It lies buried in the sand, with its great mouth open. A relative, the lesser sand eel, when frightened dives into what seems like an opening in the sand. The result is that the greater sand eel is nearly always found with a lesser sand eel, head down, in its stomach.

The ways of fish are being studied with the possibility of finding something human fishermen have not yet thought about. Thus far nothing strikingly new has developed. There recently has been much interest, says the report, in "electric fishing—either stunning fish or directing them into nets by means of electric currents." But, it is pointed out, "the fishes themselves have long ago adopted this for their own use." The electric ray on each side of its flat, round body has an area in which numerous cells are modified to produce electricity. This is not really so amazing when we consider that electrical impulses are generated normally in small amounts by both nerve and muscle cells. In these particular fishes, however, the electrical impulses are considerable and the arrangement of cells, like those of a battery, builds up a total electric potential sufficient to stun or even kill smaller animals in the surrounding water.

In only one case has man been able to use fish to catch fish. This has been by means of the remora, or sucking fish, which has the habit of attaching itself by means of suckers to other fishes. In 1494 Columbus witnessed the use of a captive remora for capturing turtles. It still is used for this purpose in parts of Australia and China.

The sucker fish has quite strong powers of adhesion. In the ordinary course of its life it attaches itself to sharks or other large fishes and enjoys a free ride until it comes across food. When used for fishing, it is fastened with a line around its tail and tethered to the canoe. The native paddles as close as possible to the intended victim without disturbing it. The remora then is thrown into the water toward the turtle, to which it automatically attaches itself. Once the remora is securely fixed to the turtle, the fisherman carefully plays his light line until the reptile is brought into the boat. This must be done with care because of the diving habits of turtles. They are likely to run away with lines, sucker fishes and all.

WORMS THAT ARE FLOWERS

There are carnation worms and chrysanthemum worms. There are fairy gardens of worm asters and cornflowers at the bottom of the sea. Pink, red, purple, green, and yellow petals are tentacles of worms whose tube-encased bodies, stems of the flowers animals, are buried in inshore bottom ooze or mud-filled rock crevices.

Among these worms are masons and architects that build the houses in which they pass their lives brick by brick and pebble by pebble, with an exquisite craftsmanship hardly rivaled among animals. The blossoms and architecture have, so far as known, no utilitarian function. Nature is a painter and a poet. Forever she probes with intellect, instinct, and emotion to capture fleeting fragments of colors, lights, and harmonies of the ineffable which can be woven into the material garments of life. Among her notable successes are the sabellids and serpulids and terefillids. They are tube-dwellers—thus distinguished from their free-wandering kin— polychaetes such as the fearsome Aphrodites. Many of them have been given the names of the golden-haired nymphs who, mounted on sea horses, formed the retinue of Poseidon in mythology. Loveliest of these nymphs was Amphitrite, who became the bride of the sea god and queen of the coral-forested deep. Quite appropriately, among the fairest of the sabellids is the amphitrite, essentially world-wide in distribution.

These worms are especially facile as builders. One, for example, makes the brick with which it erects the cylindrical house that is its home for life. Extending from its head are sixteen tentacles, eight on each side, fringed with petal-like outgrowths. These tentacles are joined by membranes at the base so that, when extended, they have the appearance of two fans. When the fans are brought in contact, they form a funnel with which the animal collects mud. At the bottom of this funnel is "a singular organ by which the mud, mixed with a cement-like secretion of the worm itself, is moulded into pellets. These pellets are laid, one by one, like bricks, to form the walls of a flexible tube from twelve to fifteen inches long and about as thick as a goose quill."

This particular British sea worm, Amphitrite ventilabrum, is almost as notable for the beauty of its blossom as for its masonry. Each of the tentacles has about a thousand of the petal-like processes and each of these, it is claimed, is capable of some degree of independent action. "It is no exaggeration to affirm," wrote the eighteenth-century British biologist Sir John Dalyell, "that the will of this lowly, defenseless creature is fulfilled by control of at least twenty thousand living parts."

The color of the petals is basically straw-yellow, dotted and banded with brown, rouge, red, and green. "While dredging in the river Roach," Dalyell reported, "I have come upon banks where these worms existed in hundreds of thousands and appear in masses of large extent growing erect like standing fields of corn."

Of another British tube builder which builds tubes of cemented shells or pebbles near the roots of large sea weeds, Rev. Richard Johnston says: "Sabellarid angilica is a timid, lively, active creature whose most prominent ability is that of constructing a dwelling for itself from sand grains. It is firm, durable, and capable of great resistance. They are not easily crushed. Some appear much more brittle. Most of the dwellings are lined with a soft, silky substance formed of exudations from the body. The worms have a great preference in building materials. They always prefer sand or shells. Powdered glass is used reluctantly and soon rejected. Some tubes are short and confined, others considerably prolonged so as to afford safe retreats in danger. Some architects seem to persist in prolonging the fabric as long as material can be found. They never weary of working. Grains of sand are selected and adopted for precise spots and gelatinous matter secures them in the tube walls."

Perhaps the most notable of all the worm builders is a five-inch-long species found in South African waters, pectinaris capensis, described by Sir

John McIntosh: "The beautiful straight tube formed by this animal was composed of the spicules of sponges in short lengths placed traversely and fixed by secretion so as to form a perfectly round tunnel gently tapered from the wide to the narrow end. The spicules appeared of the same size throughout the tube. The inner surface was as smoothly formed as the outer. The labor involved in selecting and fitting with such marvelous skill the sponge spicules composing so large a tube must have been very arduous. One tube lasts the animal for life."

McIntosh tells of another South African architect worm that "builds out of grains of sand arranged in a single layer like miniature masonry and bound together by waterproof cement."

There are, however, widely differing degrees of artistry among the tube-dwelling polychaetes. Some tubes are rough, fragile, long, bent in various directions, and united in colonies several inches to a foot across. Sometimes tubes three to four inches long are attached horizontally to the undersides of rocks.

A large and singular terebellid is Amphitrite ornata—twelve to fifteen inches long with orange-brown tentacles capable of being extended eight to ten inches. These are kept in constant motion gathering food and material for building. The bodies of these worms are filled with blood, but there is no circulatory system. The blood, however, apparently can be forced into any part of the body by muscular contractions. The tentacles can be turned voluntarily in any direction by forcing blood into them.

Tube-building, flowering worms excited the wonder of Quatrefages as he observed them along the Bay of Biscay in the nineteenth century:

"On these coasts so violently beaten by waves we often observe small hillocks of sand pierced by an infinite number of minute openings. These little hillocks which look very much like thick pieces of honeycomb are in reality populous cities in which live in modest seclusion tubiculous annelids, the hermellas—(sabellarids) as curious as any that fall under the notice of the naturalist. The body, about two inches in length, is terminated in front by a bifurcated [two-forked] head bearing a bright double golden crown of strong, sharp silk threads. These brilliant crowns are not mere ornaments, but are the two sides of a solid door, or rather true portcullis, which hermetically closes the entrance to the habitation when, at the least alarm, the worm darts with the rapidity of lightning within its house of sand.

"From the edges of the head of this worm issue fifty to sixty slender, light-violet filaments which are incessantly moving about like numerous minute serpents. They are so many arms which can be lengthened or shortened at will and which, seizing the prey as it passes, bring it to the hollow, funnel-shaped mouth. On the sides of the body appear little projections from which issue bundles of sharp and cutting lances. Finally, the back is covered with cirrhi, recurved like circles, whose color varies from dark red to deep green."

Most conspicuously flowerlike among the worms are the serpulids—"little snakes."

Found the world over, they furnish passable imitations of practically all the flowers in an old-fashioned Virginia garden. Among them, for example, are the animals of inshore South African waters, described by Prof. McIntosh. Their wreaths of branchia "look like pinks, but in some varieties are purple at the base, with narrow bands of bright red and pale green. In one variety the blossoms are yellow or orange and the body is usually greenish-yellow." "The instant it is disturbed," McIntosh says, "this worm withdraws its lovely wreath into its tube and closes the aperture with a curious plug, funnel-shaped and placed at the end of a rather long pedicle."

The Rev. D. Johnston describes a British flower worm (one of the sabellids) about an inch long, whose eight-inch-long tubes grow together, attached at the bottom to a stone or abandoned shell. The tube has a silk-like lining.

"Into this tube," says Johnston, "it can withdraw with lightning-like rapidity when alarmed. Extending across its back is a row of microscopic hooks, or 14,000 to 15,000 teeth. These are used to catch the lining of the tube and draw the worm back."

The filaments which form its blossoms, he says, are comb-like, arranged in two rows, one on each side of the mouth. They form a coronet. Under low magnification each is seen as a pellucid, cartilaginous stem from one side of which springs a double series of secondary filaments through which red blood can be seen flowing.

Some of the most conspicuous flower worms are found alone: the Atlantic coast of the United States. On diving into Chesapeake Bay one encounters tiny, colored clusters of feathers that are really gills of annelid worms. They flick instantly out of sight as their owners withdraw into tubes in the rock crevices. The blossoms are bright orange, each surrounded by a white haze caused by thousands of minute tentacles straining the water for the tiny organisms upon which they feed.

From New Jersey to Cape Cod is to be found a purple-blooming serpulid with white stems of calcium carbonate three to four inches long and an eighth of an inch in diameter.

A widely distributed family related to the serpulids are the fabricinae, or "feather dusters." These animals, only a few millimeters long, live in the upper layers of mud in tidal basins. They are so thoroughly covered with slime and debris that they are likely to be completely overlooked. The body is thread-like except for the crown of tentacles, with from seventy to a hundred featherlike filaments. In some varieties these are white, in others translucent.

THE HEAVY TOLL OF BIRD MIGRATIONS

A migration that takes a toll of millions of lives takes place every year between North and South America.

Dr. Alexander Wetmore of the Smithsonian has had the experience of standing on a lonely beach on the coast of Venezuela and actually watching North American birds arrive at the end of their gruelling journey, exhausted and emaciated. Every day over his camp on the shore passed familiar birds from home—sandpipers, yellowlegs, bobolinks, barn swallows and warblers.

"There was brought to me more definitely than ever before," Dr. Wetmore reported, "the tremendous loss of life that this journey entails. The wastage of modern human battlefields, though terrific beyond words, is nothing in comparison. On this open shore small feathered migrants often made a landfall in a state of evident exhaustion. In the early morning I found little groups of them feeding on the short herbage. Some obviously had barely made a landfall after an exhausting sea journey. In some of those that I handled the flight muscles that move the wings were reduced to thin bands through which the angular ridges of the breast bones protruded. It was easy to visualize the hundreds of thousands that had wandered over the water until they fell to drown, and the hundreds of others that arrived only to succumb to the strains imposed by their exhausting journey."

DEADLY SNAKES THAT TAKE LIFE EASY

Deadliest of serpents are the Pacific sea snakes. A bite almost certainly would be fatal to a human being. Yet native children of the Palau Islands in the South Pacific play with these reptiles with complete impunity. They pick them up and toss them from one to another just as American children play "catch." Natives of the Palaus look upon the reptiles with complete indifference.

The term "sea snake" is somewhat of a misnomer. Actually the creatures spend most of their days asleep among rocks on beaches. They are excellent tree climbers and like to sun themselves in crotches of branches.

At dusk, however, they move out to the reefs where presumably they spend most of the night pursuing small fishes, their principal food. They are excellent swimmers and their bodies have been somewhat modified, with flattened, paddle-like tails, for sea life.

Fortunately, on land at least, they are sluggish and non-aggressive. They hardly can be induced to bite and will suffer almost any indignity without retaliating. About the only way a person would be likely to be bitten would be by stepping directly on the head of one of these snakes with bare feet. This is an unlikely event, for the sea snakes do not spend any time under shallow water where they would be a peril for bathers.

Some are quite beautiful, about five feet long and banded with black and white. Their capture is easy. It is simply a matter of pinning down the head with a stick and picking up the snake by the neck.

Throughout the entire sea snake area in the Pacific there are only five or six instances reported where the serpents have bitten humans. In every case the victim has died; there is no anti-venom against the sea snake toxin.

Some years ago Dr. Herbert Clark, former director of the Gorgas Memorial Laboratory, dove off a boat in Balboa harbor and swam ashore, a distance of about 200 yards. As he neared the shore there were alarmed cries from the deck he had left. Dr. Clark looked around. He found he had unwittingly swum through a school of several thousand black and white serpents, each about two feet long. None had touched him.

WEIRD PLANT-ANIMALS

Near the bottom of life's pyramid there is a weird race of plant-animals. They are among the closest of all many-celled living things to the primaeval protoplasm from which all life arose.

They are the slime molds found on decaying logs and tree stumps in damp woods or on piles of rain-soaked dead leaves in shady gardens. The nightmarish mycetozoa—botanists call them myxomycetes—are timeless survivals out of living creation's dank, warm cradle. Some of the weirdest imaginings of malevolent life on other planets picture it in the form of gigantic slime mold aggregations—undifferentiated masses of naked protoplasm endowed with a malign intelligence which has evolved without the intermediaries of nervous systems or brains.

These organisms can be considered one of nature's probing experiments towards higher forms of life. The experiment was a failure, but unlike most of nature's discards these organisms have survived. Even now they may be engaged in a process of evolution all their own.

Biologists are not entirely agreed in which kingdom to place the organisms, although they usually are classified with the plants. They start life as spores, like the dust of molds or toadstools whose single-celled particles serve the same reproductive function as seeds in higher plants. From each spore arises from one to four animal-like organisms, hardly distinguishable from the one-celled protozoan animal, the amoeba. Each swims about freely for a time by means of tentacle-like arms, the flagellae.

These free-moving living particles are known as "swarm cells". Each is an individual with a film-like skin separating it from all other individuals. That is, the protoplasm of each cell is enclosed within a boundary and in the center of each is a nucleus. These one-celled "animals" wander about freely for a few days. During this time they may mate, as individuals. More commonly each loses its flagellae and splits into several fragments. Each of these fragments becomes a complete organism. These mate, with complete fusion of their bodies. The result is a double plant or animal—depending on whether it is observed by a botanist or zoologist—known as a zygote. The fragments are extremely voracious little creatures devouring greedily the one-celled plants, or bacteria, which they encounter.

When the fusion is complete the zygote, in turn, starts to split up into single-celled organisms but after a few divisions hundreds of these single-celled animals coalesce into a tiny ball, like the seed pod of a plant. In a few days thousands of these spheroids collect into a so-called "plasmodium". The hitherto individual pseudo-protozoans meanwhile have lost their cell walls. The primaeval substance of millions is mixed together into a slimy mass full of cell nuclei. This is an aggregation of "naked protoplasm". It is hardly to be compared with the body of any higher plant or animal where each cell retains something of its individuality, however closely its activities may be coordinated with those of its fellows in the same community. The mass proceeds to behave like a voracious animal. It moves and feeds as a unit and apparently with a purpose. Within the naked protoplasm there is apparently some incomprehensible sense of fellowship which eventually evolves into consciousness and intelligence, developing nerve and brain on the way upwards. It would be hazardous to say that this evolution could have taken no other path.

From the central body great numbers of thread-like filaments are sent out to penetrate the substance of rotting wood or the surface of a dead leaf. These threads seem to be like an army's scouting parties, pushed ahead to locate supplies when advancing troops are living off the country. When a supply is found they are drawn in and the whole slimy organism acts once more as a coordinated whole.

The plasmodium moves forward steadily for about 50 to 60 seconds, pauses for a few moments, and then reverses itself and creeps backward, but never quite so far as it previously had gone ahead. Then, after another pause, it crawls forward again. Thus there is an overall slow advance and at the bottom of life the slime molds lay down the pattern of progress recapitulated in human societies and civilizations as well as in the lives of individual men and women. They merit consideration in the philosophy of history.

The advancing mass of raw protoplasm acts like an animal and grows like an animal as it ingests food, with constant splitting of the cell nuclei which it contains. There are vacuoles within the protoplasm in which the food particles are ingested. They then are digested by means of enzymes (body chemicals), as in higher animals.

Such a plasmodium can be taken from its damp habitat and dried. Then it will roll up into a ball and pass into a resting stage from which it will revive completely in a few hours when supplied with moisture again. The ball may keep its vitality for several years.

Some species pass as much as a year in the active plasmodium stage, and some a few days. At the end of this phase of its existence the mass of raw protoplasm breaks up into fragments—sometimes as many as a hundred. Then, as the process is described for one common species "in an hour or two each of these fragments has risen into a pear-shaped body with a narrow base, a dark stalk being just apparent through the translucent white substance." In about six hours the black, hair-like stalk has grown to its full length and bears at its top a young "sporangium" consisting of a globule of viscous plasma with a diameter about a fifth the length of the stalk. This globe is about the size of a mustard seed and ranges in color from pure white through golden-yellow, light crimson, violet, purple and black.

A pink flush now begins to pervade the sporangium caused by the formation of branching threads. The nuclei in the plasma still present the same appearance as those observed in the streaming plasmodium. In about another hour these nuclei show the beginning of division. As this process develops the plasma becomes separated in masses of two spores capacity. An hour later the nuclei have divided and the young spores are forming. Their color rapidly changes. In about the first twenty hours after the first concentration of the fragments of the plasmodium they have matured and present the appearance of minute black pins standing in regular order on wood. The ripe fruit, or sporangium, then dries and breaks.

On placing the spore in water its membranous wall slips off and the naked contents lie for several hours without apparent change in an ellipsoid form. Constriction then takes place and the ellipsoid splits into one to four globular bodies adhering together and exhibiting slow amoeboid movements. Each globular body now develops a flagellum—a long, whip-like extension, and the cluster swims away by means of these flagellae.

Now the whole life process is ready to be repeated. There are more than 400 species of these slime molds and they are distributed over all the temperate and tropic zones. If only the spores and the stalked little ball containing them are considered, the slime mold would be placed squarely in the kingdom of plants. But when the protoplasm escapes from the spore and starts moving about ingesting bacteria, the behavior is that of a one-celled animal. When the cells unite to form a plasmodium there is a close likeness to a many-celled animal.

WEIRD WAYS OF BIRDS

Among the most fantastic forms of animal behavior is that of the honey guides, African birds distantly related to the American woodpeckers. They "guide" men, baboons and ratels to the nests of wild honeybees—supposedly so that these nests will be broken open.

Throughout the three centuries since the unusual behavior of the bird was first reported by a Portuguese missionary it has been the subject of many fantastic accounts, some of which attribute a far higher degree of intelligence to the birds than they possibly could possess.

A long-continued study of this behavior has been made by Dr. Herbert Friedmann, Smithsonian curator of birds. Dr. Friedmann himself has observed at least 23 instances of the habit and has collected much other well authenticated data from African associates. He describes the behavior from his own observations:

"When the bird is ready to begin guiding it comes to a person and starts a repetitive series of churring notes, or it stays where it is and begins calling these notes and waits for the human to approach it more closely. These churring notes are very similar to the sound made by shaking a partly full, small matchbox rapidly sidewise. If the bird comes to the person it flies 15 or 20 feet from him, calling constantly and fanning its tail.

"It usually perches on a fairly conspicuous branch, churring rapidly, fanning its tail, and ruffing its wings so that at times its yellow shoulder bands are visible.

"As the person comes to within 15 to 50 feet the bird flies off with a conspicuous initial downward dip, and then goes off to another tree, not necessarily in sight of the follower, in fact more often out of sight than not. Then it waits there, churring loudly until the follower again nears it, when the action is repeated. This goes on until the vicinity of the bees' nest is reached. It waits there for the follower to open the hive and usually until the person has departed with his loot of honeycomb, when it comes down to the plundered bee's nest and begins to feed on the bits of comb left strewn about. The time during which the bird may wait quietly may vary from a few minutes to well over an hour and a half."

African natives regard the bird as an almost infallible guide to honey. They try to attract it by grunting like a ratel or chopping on trees to imitate the sound of opening a nest. The habit is apparently instinctive; it presumably originated before human beings appeared, perhaps starting with the ratel or some of its honey-eating ancestors.

Curiously enough, the honey bird does not seem interested in the honey, per se, or in the grubs of bees found in the nests. It has an insatiable appetite for the wax, which it will take wherever it can be found. The first account of the bird was of an individual which fed on the wax candles of a church. It appears to have a peculiar ability to digest wax presumably to extract the nutritive elements contained.

THE FANTASTIC SEA HORSE

A fish with the head of a Lilliputian horse, the tail of a monkey, the shell of a beetle and the pouch of a kangaroo...a creature that reverses the ordinary course of nature in that "child bearing" is exclusively a function of the male....Perhaps in no other animal have been packed so many anomalies as in the little hippocampus, popularly known as the "sea horse".

These weird creatures are almost world-wide in their distribution through ocean waters where there are growths of sea vegetation. They have provided the models for some of the monsters of human nightmares. Actually they are small, feeble, almost defenseless creatures.

The head unquestionably is similar to that of a miniature horse in general outline. The neck, however, is not a neck at all. Fishes have no necks and hippocampus is no exception. What looks like a neck is the upper part of its abdomen, considerably contracted.

The body is covered with a jointed, chitinous shell, like many of the insects. This peculiarity left early naturalists in doubt as to whether it actually was a fish or some sort of monstrous water bug. It is, of course, a true fish with no insect affiliations. The hard shell makes it a feeble, inefficient swimmer. It is able, in fact, to swim at all only because of a large air bladder so delicately adjusted to the specific gravity of the animal that if a gas bubble the size of a pinhead is let out by a puncture the sea horse sinks to the bottom. There it can only crawl about clumsily until the wound is healed.

Because it is so poor a swimmer the hippocampus must have other means of adjustment to its salt water environment. This is afforded by a prehensile tail which it can wrap around the stems of water plants. This kind of a tail is found among a few mammals, notably the smaller monkeys. So far as is known, no other fish has anything of the sort. The animal is most frequently observed in a state of rest, its tail wrapped around a plant and its body standing nearly erect in the water.

Its food consists of tiny crustaceans and other sea organisms of like size. Because of its poor powers of locomotion, it must wait for those which come within reach of its jaws which work with lightning-like speed, or for those which will wait accommodatingly for it to come and get them.

Hippocampus can move its eyes independently of each other, thus looking backward and forward at the same time. It would be rather difficult for a predaceous organism to take it by surprise, but on the other hand it would have little ability to fight back or flee if attacked. Some species, at least, have considerable ability to change color to blend with the environment. Bright red, pink or yellow specimens when caught fade rapidly to normal mottled gray.

Probably the greatest anomaly of the hippocampus family is its way of reproducing the species. The male actually "gives birth" to living young. The process, so far as known, is unduplicated in nature. Unfertilized eggs are laid by the female. She places them, a few at a time, into a pouch-like organ on the underside of the male's body. In some fashion still unknown to biologists they are fertilized in the transfer. Within this pouch the eggs are incubated and there the young remain for several days after they are hatched. Then, fully equipped to take care of themselves, they are expelled into the water. So far as has been observed, there is no further parental interest in them. This male pouch might be considered as filling the double function of the womb of a placental mammal and the pouch of a marsupial like the kangaroo.

The sea horse also has the distinction of being one of the species of fish that "talk". In recent years "talking fish" have become a matter of considerable interest to the Navy because of the confusion they cause in the interpretation of underwater sounds. They give every indication of talking to each other. They produce loud clicks similar to the snapping of a finger. These also have been compared to the clicks of a telegraph. They were especially notable when an animal was first placed in the tank and apparently was confused by the new environment. It would cruise back and forth across the container, standing upright and its prehensile tail curled over its back, emitting the characteristic sounds at intervals of from a half to three quarters of an hour.

When two sea horses were kept in separate jars adjacent to each other in an experiment it appeared as if they were trying to converse. First one would emit a series of clicks. Then the other would answer. The sounds are produced by snapping the jaws together. In nature these probably are mating calls.

THE GREAT SEAL MIGRATIONS

The great annual northward migration of the seals is one of the most remarkable phenomena of animal life. It seems to be without organization and without leadership, yet toward the end of March each year the hundreds of thousands of cow seals and pups scattered over thousands of square miles of water start at about the same time in three great groups bound for three specific places. It has been the same for centuries, perhaps millenia. Each animal moves at about the same rate so that all arrive within a few days of each other. They do not move in compact masses, like birds.

The American herd of about 1,500,000 is by far the largest of the three. It goes straight to the Pribiloff Islands where it goes ashore on two almost barren islands—St. Paul and St. George. The Japanese herd, numbering about 40,000, makes for Robben Island, off northern Japan. The Russian herd, now estimated at about 200,000, goes to a few rocky islands of the Commander archipelago, off Kamchatka.

The moving herds consist almost entirely of females and young. The bulls winter further north, tend to be solitary during the winter, and precede the cows to the summer homes. The breeding season lasts for about two months. During this time the bull never eats or touches a drop of water. He never leaves the land. He arrives sleek and fat from the ocean pasture and is able to survive entirely on stored energy. This keeps him alive, even when he fights scores of terrible battles with younger rivals. Towards the end of summer he naturally is a sorry-looking creature.

One day, actuated by some common impulse, cows and calves depart. Then the bulls, their arduous labors of race propagation over for ten months, draw back among the rocks and spend two or three days in sound sleep before returning to the sea to replenish themselves.

Cows have very little reserve energy and must return to the water every two or three days, leaving their nursing pups ashore. On her return from one of these feeding expeditions, a cow goes straight to her own pup among the thousands on the rocky beach. Presumably she locates it by the odor. Few animals grow more rapidly than the seal pup. Within a few weeks after birth it is almost as large as its mother. This is an essential provision of nature, for it must have sufficient size and strength to care for itself in the open sea, once the southward migration starts. It is fully the size of the mother when it comes back the next year. There is an old idea that seal pups must be taught to swim. This is denied by government observers at the Pribiloff breeding grounds. When thrown into the water for the first time they swim ashore without difficulty. They will not, however, venture into the sea voluntarily but must be pushed off the rocks by the mothers.

St. George and St. Paul islands are the only two spots under the American flag, except for certain atomic energy and military installations, which are absolutely barred to visitors without special government permits. These, as a rule, are given only to scientists studying the behavior of the seals. On each island there is an Aleut village whose inhabitants attend to the butchering of the animals each summer. This is confined entirely to three-year-old males who congregate by themselves. The only other killing permitted is by Aleuts along the coast for whom sealing is the traditional means of livelihood, but this now is so restricted that the annual toll is very small. The sealing must be done from an open boat, use of firearms is prohibited, and the Aleuts cannot be under contract to furnish skins.

MONSTERS WITH BUZZ SAWS

"But if, retaining sense and sight, we could shrink into living atoms and plunge under water, of what a world of wonder would we form part. We would find this fairy kingdom peopled with the strangest creatures—creatures that swim with their hair, have ruby eyes blazing deep in their necks, with telescopic limbs that now are withdrawn wholly into their bodies and now stretched out to many times their own length. Here are some riding at anchor, moored by delicate threads spun out from their own toes. There are others flashing in glass armor, bristling with sharp spikes or ornamented with bosses and flowing curves; while fastened to a green stem is an animal convulvulus that by some invisible power draws a never-ceasing stream of victims into its gaping cup and tears them to death with hooked jaws deep down in its own body."—*The Rotifera* by C. T. Hudson and P. H. Goose, London, 1886.

The rotifers or wheel animalcules are fantastic creatures. They were first seen by the Dutchman Antonius van Leeuwenhoek, credited with being the inventor of the microscope. "On the 25th of August," he wrote to the Royal Society of London with which group of savants patronized by Charles the Second he was in regular correspondence, "I saw in a leaden gutter on the front of the house for a length of five feet some rain water had been standing which had a red color. It occurred to me that this redness might be caused by red animalcules. I took a drop or two of the water and looked at it under the microscope."

He found a confusion of "red-eyed monsters armed with teeth like those of the balance wheel of a watch, which appear to be projecting forward towards the head. They seem to whirl around with a very considerable velocity, by which means a rapid current of water is brought from a distance to the mouth of the creature who thereby is supplied with many invisible food particles."

This discovery is of considerable significance in scientific history because, more than any of his previous findings, it caused the Amsterdam spectacle-maker to question the then widely held belief in the spontaneous generation of living things.

"They can," he wrote the Royal Society in 1774, "continue many months out of water and be dry as dust, in which condition their shape is globular, the bigness exceeds not a grain of sand, and no signs of life appear. Notwithstanding, being put in water, the globule turns itself about, lengthens by slow degrees, becomes in the form of a lively maggot, and most commonly in a few minutes afterwards puts out its wheels and sweeps the water in search of food. But sometimes it may remain a long time in the maggot form and not show its wheels at all."

Such tiny organisms capable of such long periods of suspended animation, Leeuwenhoek held, could be blown by the wind for long distances. Thus the sudden appearance of living animals in supposedly lifeless water did not indicate they had been born or created there.

The microscope designer had found, moreover, an hitherto unknown race, giants of the microscopic world and among the most fantastic of all animals—the rotifers.

These usually invisible animals with buzz-saws on their heads—the largest not more than a quarter-inch long and the majority less than a twentieth—seem to have gone further beyond life's normally accepted frontiers than any other animals. One species lives comfortably in hot springs where temperatures go above 120 Fahrenheit. Others can be frozen in solid cakes of ice for weeks and show no ill effects. Sudden changes in temperature, however, often are fatal. On tops of Antarctic mountains projecting out of ice two miles thick, the little rotifers are found among sparse growths of lichens, the only animal life which approaches closely to the South Pole on land. There is no reason why they should not thrive in the hardly less hospitable mountains of Mars. They might have been carried there in light propelled earthdust.

The majority are fresh-water creatures. A few live in damp moss and a few species have obtained a foothold in the sea. Some live in immense colonies, permanently attached to stones. Some are free-living individualists who crawl like leeches, or swim rapidly. Some are parasites in the cells of water plants or in the gills of fresh water crabs. Others cling to floating plants or to water animals, to be carried from place to place. One highly social group lives in free-moving communities of forty or more individuals, attached to each other by their tail ends and radiating from a common center like wheel spokes. The usual color is reddish and most rotifers have one or more glittering red eyes. In a few cases these eyes are inside the bodies of transparent species.

Despite their minuteness, these predatory giants of the world invisible are highly developed animals. Each has a body divided, like that of a mammal, into three major segments—head, trunk, and extremities. In some the skin is hardened into an armor-like covering. Some have a panoply of defensive spines and bristles.

Inside the skin is a cavity full of watery fluid—it contains no corpuscles like blood—in which float the more important vital organs. In most animals there is tissue of some sort in which nerves, muscles, and glands are imbedded. In rotifers, however, there is very little of this connective tissue. Under a microscope one generally can see with some clearness each individual cell. These cells can be counted, for at the most there are only a few thousands, compared to the millions of millions that make up the bodies of larger animals. The muscles are not banded together, but consist of isolated strands whose job is to pull the head inside the armored trunk when faced with any threat, and to bend the body in various directions.

All rotifers have two organs unique to their race. First is the "buzz saw". This is a crown of tentacles, quite similar in appearance under low magnification to a circular saw, which is constantly whirling. Its purpose is to create eddies in the water which will bring food particles to the mouth, a funnel-shaped opening on top of the head. In free-living species the saw may have some function as a propeller.

Second is the mastax, or "chewing stomach". Every rotifer has two stomachs, one for masticating and one for digesting. The mouth opens directly into the first. It is provided with two horny, serrated jaws which crush toward each other and tear to bits the minute animals and plants which are the creature's food. The jaws are provided with several hard parts, adapted for biting, crushing, holding, and tearing.

In the permanently anchored rotifers the rear of the body is prolonged into a stalk from the end of which a cement-like substance is secreted. This permanently attaches the animal to something, usually a stone. In some of the free-living forms the "foot" is replaced by one to twelve "leaping spines" by means of which the owner can spring suddenly forward several times its own length to capture an unsuspecting victim. This is most often some floating one-celled creature of the water-drop jungle, such as a protozoan elephant.

The male rotifer is usually much smaller than the female—sometimes nothing more than an appendage she carries about with her. The fantastic worlds of all sorts of rotifers are predominantly feminine worlds. For some species, in fact, males never have been found, but there is little doubt that they exist.

TWO-HEADED SNAKES AREN'T RARE

Two-headed snakes probably are quite common. About 200 cases have been reported. Dr. Bert Cunningham of Duke University, who has studied several living specimens, has this to report about such snakes: "The heads play together, fight over a morsel of food even though it will go into the same stomach through either mouth, attempt to swallow one another, and sometimes fight fatal duels. Each head has a brain of its own. Few grow to any size. In this case two heads are not better than one, especially when they disagree when a second means escape or death."

FANTASTIC SEA CREATURES

Coral-forested waters around the Gilbert and Mariana Islands in the Pacific are yielding some of the most fantastic sea creatures known to science.

Extensive collections have been made since the war by Dr. Leonard P. Schultz, Smithsonian curator of fishes. Notable in the collections are snake, worm and moray eels, all bottom dwellers in tropical waters. Snake eels are, as the name indicates, superficially almost indistinguishable from serpents. On their tails they have hard points which are used as drills. They burrow straight downward in the bottom sand, tails first, until only the heads protrude above the surface. The worm eels belong to the same general group but are much smaller and slenderer—about the diameter of a lead pencil and reaching lengths up to two feet. Larger worm eels have been reported.

Both these groups consist of relatively timid, inoffensive creatures. Far different are the moray eels, members of a closely related family. They are as much as ten feet long, have razor-like teeth, and are described by Dr. Schultz as about the most vicious creatures in the sea. In disposition they probably are worse than the worst sharks and easily can bite through a man's hand.

Probably the most poisonous creature in the collection is a variety of sting ray, weighing about 200 pounds, which was speared at the bottom of 20 feet of water. This animal, like all stingarees, has a tail armed with long, poisonous barbs. The venom could be lethal to a man. After it was speared, the ray remained very much alive and the problem of bringing it to the surface was difficult. This finally was accomplished by two of Dr. Schultz' collaborators. First one would dive, grasp the handle of the spear, and lift the creature a few feet, always holding it far enough away to be clear of the barbs. After the first man became exhausted, the other would relieve him while he came up for air. Thus the specimen finally was gotten on board through a series of relays.

Curiosities of the collection are the cardinal fishes—brilliant red, very active, and including some of the smallest marine fishes. A few species attain full growth at about three-fourths of an inch. These are the most notable of the "mouth breeders." The female lays the eggs and the male carries them in his mouth until they hatch. Inch-long males sometimes carry as many as 400 eggs, nearly all of which hatch.

Other curiosities are the pipe fishes, hard-shelled animals which look like bits of small, segmented pipe. They range from two inches to a foot long and are related to the more familiar sea horses of temperate waters. They are sluggish burrowers in coral reefs. As among sea horses, the male gives birth to the young. The eggs are deposited in pouches on the male's belly where they are carried until they hatch.

THE VARIETIES OF RAVEN LANGUAGE

While "nevermore" apparently is not in the vocabulary of the raven this big black bird of the wilder parts of the country has a considerable variety of sounds nearly as ominous.

Raven "language" has been intensively studied by the noted ornithologist, Dr. Arthur Cleveland Bent. Citing various bird observers, he lists the following calls:

A distinct, hollow, sepulchral laugh, haw-haw-haw-haw, which may be heard at almost any time.

A series of "crawks" sounded while on the wing, interspersed with a musical note that sounds like ge-lick-ge-lee.

A strange call like thing-thung-thung which is similar to the mellow twang of a tuning fork.

Another expression has a metallic, liquid-like quality similar to the song of the red-winged blackbird, although greatly magnified in volume.

Ravens have a large range of notes from the melancholy croaks with which they chiefly are associated to striking imitations of other birds, such as geese and gulls. One of these birds will talk to itself for hours with a curious gargling sound. He becomes so absorbed in his own conversation that it often is not difficult to steal up on him during such a soliloquy.

"The raven," Dr. Bent observes, "is one of our most sagacious birds—crafty, resourceful, adaptable, and quick to profit by experience. Throughout most of its range it is exceeding shy and wary. It is almost impossible to get within gunshot of one in the open. Yet it knows full well where and when it is safe. About northern villages, where it is appreciated as a scavenger and seldom molested, it is as tame as any barnyard bird." This is especially true in Greenland where ravens infest American air bases.

Although in the north the raven frequents the seacoast and villages, from Pennsylvania southward it is entirely a mountain bird, usually living above 3,000 feet. From these heights the birds sometimes descend to the valleys, or even the islands along the coast, to forage among the colonies of sea birds. Most of them prefer to dwell among rocks and resort to perpendicular cliffs and to escarpments thrust above forests on the flanks of mountains.

WORMS WITH HYPODERMIC NEEDLES

Despite their microscopic size, nematodes (soil worms), are highly organized animals. They have muscles, quite specialized organs for feeding, a digestive system, a nervous system with a brain, and a well-developed reproductive system. Sexes are clearly differentiated. The creatures have evolved a long way from the primeval worm.

Eggs may be deposited in the soil, or in the plant on which the nematode feeds. In these eggs the immature forms, the larvae, develop and eventually hatch. If appropriate plants are available, they may begin to feed immediately. They develop through several distinct stages. At the end of each of these cycles a moult occurs.

Many of the forms which have been studied closely have a minimum life cycle, from egg to egg-laying female, of several days to several weeks. The maximum duration of life, however, may be much longer, since sexual maturity is not reached until the nematode begins to feed on the living plant. Up to this time it remains in the larval stage and lives on a reserve food supply originally derived from the egg. The time this reserve lasts depends on circumstances. In damp, warm soil the nematode will be very active and use it up in a few weeks. In cool or dry soil the supply lasts much longer, and can extend to many years.

The little worm's life is a perpetual struggle for existence. It has many enemies in the soil—insects, fungi, and other free-living nematodes. Certain of the soil fungi have "traps" especially designed to catch nematodes. Some of these are shaped like loops which are pulled tight as the worm starts to crawl through. Others are sticky surfaces on which the victims are captured, like flies on flypaper. In either case, the fungus grows into the body of the worm and kills it.

Nevertheless, the nematode population is never in any great danger of extermination. A single female root knot nematode will produce about 300 eggs in a couple of weeks. Allowing four weeks for a generation, and assuming half the offspring are females, this implies a theoretically possible fifty trillion individuals at the end of the four generations of a single summer.

Practically all roots are attacked by some kind of nematode, but many species appear to specialize on one type of plant and will not touch a different variety, even if no other food is available. Plants immune to one species may be highly susceptible to some other. A few kinds of these worms, however, appear to eat almost anything they can find underground.

All the root-eaters have a feeding organ which is much like a hypodermic needle. This is pushed into the tissue and, it is believed, a digestive juice of some sort is injected. This liquifies and partially digests the food. Then the nematode sucks it through the needle into its mouth.

The largest of the nematodes, a parasite of whales, can reach a length of 27 feet. The smallest, a marine form, is a little more than a three-thousandth of an inch long.

THE FATAL BLACK WIDOW SPIDER

The venom of the dreaded Black Widow spider is approximately fifteen times more potent than that of the rattlesnake. The comparison has been established by determining the amounts of rattlesnake and spider venom necessary to kill rats of the same weight. The extreme toxicity of the spider becomes of considerable significance since it has been reported from every state in the Union and may be increasing in numbers on the edges of cities. Probability of being bitten, however, is slight. The black widow is a timid creature, except towards her natural prey. At the first molestation of her web she retreats quickly to her central nest and does not venture out again for hours. She makes no attempt at defense, to say nothing of aggression.

Her reputation is so bad, however, that in some cases pickers have refused to work in vineyards which she infested.

PLANTS THAT ARE ANIMATED

Among the curiosities often sold in American stores are so-called "air plants"—plants that will grow on air alone without sunshine or water. This is true, after a fashion. The "plants" actually are dried skeletons of marine animals. They belong to the group which includes the jellyfish, sea anemones and corals. Their skeletons have a striking resemblance to plants.

The species most commonly sold is sea moss or Neptune's fern, an animal abundant in the North Atlantic, especially in the English channel and the Gulf of Maine. A closely related species, the "squirrel's tail," is abundant in the eastern Pacific where its silvery colonies often are washed ashore by storms. Dry beach material of these colonies is easily collected, dyed and sold as Christmas decorations.

"These are colonial forms consisting of thousands of individual animals," according to the Chicago Museum of Natural History. "Colonies of two species of sea squirrel may be twelve inches or more long. Those of some species may be several feet in length. Usually they are attached to rocks or other substrata by a rootlike base, from which spring the delicate branched stems bearing hundreds of minute polyps.

"Most of these are hydranths (feeding polyps) that capture microscopic organisms. The reproductive polyps are less common, usually larger, and different in shape. The common stem is made up of external non-cellular material, mostly yellowish or brown in color."

THE TOMATO—CINDERELLA OF VEGETABLES

A remarkable chapter in the history of agriculture is the story of the tomato which now constitutes one of this country's major crops. It appears to have first been used as a food by the Aztecs. It was introduced into Spain early in the 16th century and a century later was grown widely in England as an ornamental plant. Not until the next century, however, did it have any standing as a food. It was known as the "love apple" and was considered mildly poisonous. Folks ate one now and then on "dares."

Then it caught on as a food in Italy and by the start of the 19th century was being grown on a field scale. So far as known, it was absent from the gardens of Colonial America, unless as a rare ornamental plant. Not until the middle of the 19th century was it reintroduced to its native western hemisphere as a food crop. For a long time it acquired no great popularity. A few vines in the family garden were considered enough, since there was no tomato market.

A U. S. Department of Agriculture report calls the tomato "the prodigy of the vegetable world." Its present success is due in large part to the discovery of vitamins. Although used as a food for little more than a century it now is almost as widely distributed as wheat, a food plant which has been cultivated for at least 5,000 years.

Today the tomato crop covers about a half million acres in the U. S. alone. This crop consists of more than 20,000,000 bushels of fresh tomatoes and more than 300,000 tons of canned products. There are now about 150 known varieties, adapted to all sorts of purposes.

THE HOLIEST PLACE ON EARTH

The summit of Adam's Peak in south-central Ceylon, wrapped perpetually in priestly robes of grey clouds, is one of the holy places of the earth. There, through many centuries, the prayers of millions belonging to warring creeds have worn thin the curtain between the effable and the ineffable. It is a shrine of four of the world's great religions. In the rock is a depression that looks like a giant's footprint. Hindus believe it was made by snake-haired Siva, the destroyer. Moslems say it is the footprint of the first man, Adam, who was exiled to this mountaintop after he was thrown out of Paradise. Buddhists believe that it could have been made only by the great Gautama. Nestorian Christians maintain that it is a relic of the disciple Thomas, who brought the gospel of Christ into the East. To this spot, braving the road through leech-infested forests below and the perilous ascent along gale-swept ledges, have come generation after generation of devout pilgrims to voice a common prayer in different tongues through different intermediaries.

The pilgrim, standing by the footprint of Adam, looks down upon the forest-covered hills to the eastward. Over all the land spreads the grey shadow of the supernatural. Below him is one of the most imposing spectacles on earth—the middle slopes scarlet with the blossoms of dense forests of gigantic rhododendrons, the deep-blue patches of mountain lakes, and canyons which no human has entered—their mysterious depths hidden by wind-tossed fog. Great waterfalls roar over vine-covered cliffs. Strange sounds arise from jungles of white-stemmed palms. It is a wild land of ghosts and demons watched over by the holy mountains.

In this unearthly country native legend from ancient days has placed, most appropriately, the death valley of the elephants. There, in a pleasant hollow beside a lake of clear water—reached only by a narrow pass with high walled precipices on either side—these animals make their way from all over the island when they feel the chill drowsiness of approaching death. It has been an interminable procession of the doomed since time began. To the stricken old elephant, the coming of death brings an irresistible nostalgia which draws his faltering feet homeward to this mist-shrouded valley piled high with the white bones of his ancestors. It is his haven of rest from the weariness and disillusion of living.

The belief has deep roots in the ancient folk-lore of Ceylon. It has spread all over the East. It is embodied in the Arabian Nights. No man ever has entered this vale of death since Sinbad the Sailor, who was carried there in the trunk of a huge elephant after he had been knocked senseless when the tree in which he was hiding was uprooted by a herd of the animals. Sinbad at last found himself in this valley piled high with bones and knew that he was in the long-sought death place of the elephants.

Another Ceylon elephant cemetery is concealed in a dense forest near the ancient sacred city of Anardhupara. It is so well hidden that no man knows its exact location, although all know that it exists. Unless there are such cemeteries, the natives ask, what becomes of the remains of dead elephants?

The death of the jungle elephant remains a fantastic mystery. No very serious efforts have been made to provide a solution. Remains of these creatures that have died natural deaths seldom have been found, either in Asia or Africa. Yet obviously the great beasts are mortal, subject to various fatal ailments and to the inevitable decay of age. Evidently when one of them feels death approaching it retires to a place of the dead where it quietly breathes its last and adds its bones to those of the vast multitudes of its race that have gone before it into the unknown.

The belief is so strong that there has been a persistent search for these elephant Golgothas for the past century. Such a discovery, especially in Africa, probably would mean inestimable wealth in ivory. But, except for one or two questionable instances cited below, nobody ever has found such a place. Natives sometimes claim to know an approximate location from tradition, although they never have seen it.

Zoologists naturally frown upon the idea because of its very weirdness. They explain that the remains of very few tropical animals ever are found and that the elephant, for all its bulk, need be considered no great exception. Vultures, jackals, hyenas and other carrion eaters soon would tear the flesh from the bones. Insects would bear away the fragments they left. Jungle vegetation rapidly would cover and hide the naked skeleton.

Some credence is given to the native belief by Lieut. Col. Gordon Casserly of the British army. A persistent elephant hunter during years of service in India, he never came upon the carcass or bones of one of these animals which had met a natural death. "The idea of a vast death place of these modern mammoths hidden in the remote recesses of the Himalayas," he states, "did not seem a far-fetched one to me when I lived in the shadow of those mighty mountains and heard at night the great elephant troops pass by the little outpost that I commanded on the frontier of Bhutan, as they clamber up towards the snow-clad peaks from the forest below."

The British elephant hunter W. D. M. Bell once thought he had found one of East Africa's elephant cemeteries in the country north of Lake Rudolph. He had followed an elephant path to a grassy plateau strewn with skulls and other elephant bones, some partially buried. None of the remains, however, were recent. Bell tasted the green water of a nearby pool and found it bitter with natron. The indications were that large numbers of elephants had been driven to this pool to drink during a time of drought and had been poisoned by the water.

Maj. P. H. G. Powell-Cotton tells of finding another spot strewn with bones in the same general region which might answer the specification for an "elephant graveyard." "Here I was surprised," he reported, "to find the whole countryside scattered with remains, the fitful sun lighting up glistening bones in every direction. In all my journeyings through elephant country I do not think I have ever come across before a skeleton of one of these beasts for whose death the guides could not account. My guide called this place 'The-place-where-the-elephants-come-to-die' and assured me that when the elephants fell sick they would come deliberately for long distances to lay their bones in this spot. I had heard of these cemeteries from Swahili traders who told me they had occasionally found more ivory than they could carry. The place was well known to the Turkana, who regularly visited it to carry off the tusks."

THE VANISHING GOLDEN CARPET

The rarest plant in North America, found only four times by botanists, is a ground-hugging desert flower—the gold carpet. The plant appears, on rare occasions, in California's Death Valley. Its appearance is that of a rosette of yellow leaves, sometimes as much as ten inches in diameter, lying flat on the ground. From this rosette arise innumerable tiny golden yellow blossoms, so that the whole seems like a patch of golden carpet in the brown desert. The reason for its rare occurrence is that its seeds can germinate only after a good rain. Such rains are rare in its habitat.

The plants must spring up within a few days. Ordinarily, even then, they die with the increasing drought before blossoming—thus forming no seeds. In order for them to produce the seeds for another generation there must be another rain following shortly upon the first.

The seeds become buried in the desert soil and, in the course of evolution, have developed the capacity of suspended animation over a number of years. In the old days, it is probable, these seeds retained their fertility only for a single season. Now there may be several years between rains sufficient to spur them to germination, and even longer periods between double rains which will enable them to form seeds.

The strange little plant first was discovered in 1891. There were only two specimens and search failed to reveal any more. Two years later, however, at about the same place another single plant was reported. No others were revealed by an intensive search through the entire area.

In 1931 and 1932 Dr. Frederick V. Coville of the U. S. Department of Agriculture and French Gilman, a California botanist, again made an intensive search but could not find a single plant. They came to the erroneous conclusion that the plant might be native to the mountains, from which occasional seeds were washed down after heavy rains. A few years later Mr. Gilman again took up the search and succeeded in locating the plant in four places. He found 14 individuals altogether and watched their growth carefully. Only three became large enough to flower and produce seed. The others dried up and died when they had only a few leaves and no branches. Later, however, Gilman found many specimens of the gold carpet scattered over low hills in the neighborhood.

These little hills all were whitish in color. This led to the idea that the chemical composition of the soil might have something to do with the appearance of the plants. Analysis, however, showed there was no basis for this assumption.

In the distant past, the gold carpet may have been a very abundant plant, germinating and flowering annually in a reasonably moist climate. Probably a few individuals developed the capacity of producing seed which would remain fertile over a lapse of years. When the climate changed these had a decided advantage over their fellows.

Apparently the gold carpet is a plant in the process of extinction. The continued existence of the species depends on the dormancy of a sufficient number of seeds to carry it over unfavorable years of inadequate, or inappropriately timed second rains. If Death Valley becomes drier and drier and years with suitable double rains become more and more infrequent the vitality of the seeds in the soil eventually will be insufficient to span the long periods when no seeds are produced.

EVOLUTION OF THE BIRD

It's a long call from the birds with teeth that hovered over the strange world of the dying dinosaurs 150,000,000 odd years ago to the chorus of sweet singers whose music opens sleepy eyes on May mornings of the present. The long and devious road can be traced from the grotesque archaeopteryx and archaeornis—nightmare-like and long extinct flying creatures of the dawn—to the living wren and blackbird. But however complicated, the family tree of birds is simple compared to that of the reptiles or the mammals, since avian evolution has been confined within narrower lines.

Up to the time that the monster reptiles were beginning to disappear, it seems probable that all birds had teeth. Gradually, they disappeared as the group advanced into the dawn age of present life forms. First were the ancestral birds—the archaeornithes. They were essentially winged reptiles. Following them came the toothed true birds of the New World, known from very fragmentary fossil records. They included the hesperornis, the hageria and the ichthyornis. Then, representing a long advance, came creatures of the ostrich family, probably the most primitive of living birds. They are true birds but have not reached the typical modern pattern. At the top of the family tree, the highest branch of bird evolution, is the great suborder of song birds. It includes fifty families ranging from the larks to the finches and buntings.

SPEED ACE OF THE AIR

The swiftest bird flight ever recorded accurately is in the neighborhood of 175 miles an hour. Ordinary, unhurried flight averages from twenty to forty miles an hour.

The fastest flyer, according to official records, is the California duck hawk whose speed was measured with a stop watch from an airplane. Eagles apparently are much slower.

Among the more reliable bird flight speed measurements are those of herons, hawks, horned larks, ravens and shrikes. Rates range from 22 to 28 miles an hour. Flight in all these cases was normal and unhurried. Other speeds reported by the Smithsonian are: crows, 31 to 45 miles an hour; starlings, 38 to 49 miles; geese, 42 to 55 miles; ducks, 44 to 59 miles; falcons, 40 to 48 miles.

When frightened, most birds probably can nearly double their normal rate, but they cannot keep it up very long. When cruising about in search for food they fly so as not to waste their strength. This is particularly true on the great annual migrations.

Considering ten hours as a fair day's flying time over land, the measured speeds would carry crows from 310 to 450 miles between sunrise and sunset and ducks and geese from 420 to 590 miles. Considering that they fly in straight lines, this means that they make very good time from point to point. It is highly probable, however, that most migrating birds proceed in a leisurely manner and that after a flight of a few hours they pause to feed and rest.

THE REMARKABLE INSTINCTS OF THE SILK WORM

The silk worm's brain has an instinct center contained in a speck of nerve cells with a mass of less than a millionth of an ounce. This center is a microscopic so-called "mushroom body", found in both sides, or hemispheres, of the brain. The discovery, with possible far-reaching philosophical implications, came out of some of the most delicate conceivable microsurgery in which the area was destroyed almost cell by cell by means of an invisibly fine electric needle.

Doctors Carol Williams and William Van der Kloot of Harvard have made minute studies of an American silk worm, the cecropia (common along the Atlantic coast), which spins as strong and delicate threads as the Japanese or Italian domesticated silk worms. The cocoon is a marvel of apparent ingenuity, made of a single thread almost a mile long. It is made in three layers, roughly after the design of a thermos bottle. The outer layer is a tightly woven, waterproof silk bag. Inside this is a layer of loosely spun material which serves as an insulating layer. The third layer, woven around the body of the worm itself, is a bag of exceedingly fine, soft silk. Through each layer a "hatchway" is provided directly in front of the creature's head. These must be placed one in front of the other with mathematical exactitude. Through them the self imprisoned animal must escape when the time comes, and the slightest error probably would make it a prisoner forever in a coffin of its own creation.

Inside the cocoon the worm remains, adequately protected from cold and damp, for nine months. It emerges as a winged moth, whose sole function in life apparently is to lay eggs to produce more silkworms.

Spinning such a cocoon with its three quite different layers requires extreme precision of movement. Nature has not allowed for any possible variations. Yet the masterpiece obviously is not the result of any thinking, education or practice. The little worm's life span, for one thing, would not allow for any training. Every movement must be instinctive and presumably unconscious, directed by the same part of the nervous system into whose structure the pattern has been built by nature.

The house building must start at precisely the right time. Until that time, according to the Harvard physiologists, the responsible area of the brain is held in restraint by a hormone secreted from two tiny glands in the head. At the foreordained instant this inhibiting secretion ceases and the mushroom body can go into action. The spinning can be started at any time, however, by destroying the glands.

Williams and Van der Kloot tried effects of two gasses, carbon dioxide and carbon monoxide. Both acted as potent brain depressants, but in quite different ways. The first eliminated the spinning behavior entirely and permanently. The worms wandered about aimlessly, apparently trying in vain to remember what some overwhelming internal drive was pushing them to do. The automobile exhaust gas, carbon monoxide, fatal to humans but without any serious lasting effects on invertebrates because of the lack of the red cells in the blood with which it combines in higher animals, caused them to spin a worthless and meaningless flat layer of silk as long as the effect continued. When this ended the worm started to spin what remained of the mile-long thread in the customary pattern, starting from the point it normally would have reached had it not been gassed.

The biologists then resorted to their unbelievably delicate surgery. They proceeded to destroy the silk worm brain a few score cells at a time. The brain contains hundreds of thousands of cells. The destruction had no effect on the spinning behavior until they reached the mushroom body. When a few cells of this area were killed by the electric current the worm no longer could spin a cocoon but continued to wind and weave its silken thread into three flat sheets, corresponding to the three normal capsules. The weaving continued with the destruction of a few more cells, but only in a single sheet. When a few more were destroyed the entire cocoon-making behavior came to an end.

Thus, Doctors Williams and Van der Kloot concluded, they had located a physical unit of behavior. Within it was capsuled the whole "memory" of the silk worm race with respect to spinning. More than a century ago this mushroom body was discovered by the French physiologist Dujardin, who called it the "seat of instinct." At that time this was only a wild speculation on his part, without any supporting facts whatsoever.

The instinct center is found in the brains of all insects in whom group instinctive behavior has manifestation. In the honeybee worker, intellectual giant of the insect world, it reaches its greatest size. In drones and queens, who do not display much behavior of any sort, the area of the brain is quite small.

THE STRANGE WORLD OF THE SEA

Under the tossing surface of southern seas is an inferno-like realm of everlasting darkness, inhabited by multitudes of strange animals which exist almost altogether by the laws of beak and fang. Some of them are grotesque beyond the reaches of a nightmare.

Countless generations ago their ancestors, driven by hunger and competition, abandoned the familiar sun-lit world for the perpetual night of the abysmal depths. Then with each family, it was a case of survival of the fittest and variation of form and structure to fit the environment.

Here is the stark struggle for survival with the mask of sunlight, green fields and flowers discarded. It is not different in kind but in degree from the struggle that goes on continually between living things at the surface of the ocean and on the land. Down there all must eat flesh. There is no plant life intermediary between beast and beast. Plants cannot grow below the light line of the sea depths.

Out of this fierce war for existence have come creatures mostly conspicuous for their defensive and offensive equipment. Some of the fish seem to have become little more than enormous mouths with rows of long, razor-like teeth with which they seize and kill. The bodies attached to these mouths are small and slender. Such a creature is mostly head and the head is mostly mouth. Nearly all the fish carry light organs of some kind near the mouth with which other animals are probably attracted within grabbing distance.

One of the largest collections of deep sea animals was assembled a few years ago near the Puerto Rico Deep, the deepest part of the Atlantic Ocean, by a Smithsonian Institution expedition led by Dr. Paul Bartsch. This collection constituted a fair representation of the sea life at depths of about 3200 feet, nearly 2500 feet below the farthest reaches of the sun's rays. There were shrimps with long, sharp claws which fold up after the fashion of an old-fashioned straight razor. Any small creature which came within striking distance of such a razor probably would be an immediate victim. There were strange mollusks with shells like corkscrews and eels like darning needles with long, sharp beaks.

Among the most fantastic was the needle-fish. It jaws are prolonged into extraordinarily slender points, like fine needles, so that the head is nearly as long as the rest of the body—that is, about six inches. This fish was lured to the net by an electric light.

A group of flat fish, or flounders, was obtained, all of which have two eyes on one side of the head and none on the other. Instead of right eye and left eye there is upper eye and lower eye.

Other strange forms in the collection:

The hunchback fish, a creature whose strangely shaped body suggests its name.

The lance fish with long, backward-reaching spines suggestive of lances just behind the eyes.

The forceps fish, one of the most aberrant of all with its greatly extended, forceps-like jaws. There is apparently but a single genus and species in existence.

The family of snout fish with snouts almost as long as the rest of the body. At the end of the snout is a mouth.

Another strange creature taken out of the depths by this expedition was Johnsonia eriomma—the "big eye fish." Each of its two eyes is about a fifth as long as the diameter of its body. A man's eye, in the same ratio, would be about a foot long and protrude about eight inches from its socket. It also has two false eyes on its sides, near the tail. They are of the same size and approximately the same pattern as the true eyes. They probably are indistinguishable from them by other fish. They are, however, only color spots and have no visual function. They constitute a feature hitherto unknown in the fish world. The purpose of the false eyes is unknown, unless they are intended to deceive the creature's enemies. Since it is a slow-moving fish, these color spots probably create the illusion of fast movement which would fool a predatory animal of the abysses.

This fish is the second of its family ever found in the western world. The other was discovered a half century ago the genus have been found in Asiatic waters.

This eye-fish was obtained from a depth of between 150 and 300 fathoms—just about on the borderline of eternal darkness where eyes would be of no use. Fish of the depths have evolved in two directions—toward enormous eyes and toward greatly diminished ones. The first represents a struggle to see in the strange dusk. The second trend denotes giving up of a futile struggle on the part of the race. This trend is noteworthy among fish of the greater depths.

Another strange denizen of the depths is Peristedion bartschi, named in honor of Dr. Bartsch. It is an armored gurnard, of the family sometimes known as "sea robins." The shell-growing tendency among fish is largely confined to certain fresh-water catfish of South America. This creature obviously is a bottom dweller. Its entire body is covered with spiny plates which probably would make it safe from any enemy. Each plate bears a very sharp spine, about a quarter inch long. There are nearly a hundred of these on the body. This fish would probably be about the most unappetizing morsel any predatory animal ever swallowed. It is bright red.

Still another species obtained by the expedition was one of the "lantern-fish" group. These are small, minnow-like creatures who live only in the open ocean. While most fish either remain near shore or have at least an association with the bottom these are found only in deep water far from land, and never near the sea floor. Most of the millions of them in the sea doubtless live and die without any realization that there is either bottom or shore. All have rows of luminous spots along their sides which probably serve as recognition marks.

THE CANNIBAL BIRDS OF THE PACIFIC

Hordes of big black birds, about the nearest creatures imaginable to the harpies of Greek mythology, nest on desert-like South Pacific Islands. These are the vulture-like frigate birds—the Polynesian "iwas" or "thieves"—which are found by thousands in branches of the most prominent shrubs, the eight-foot-high, white flowering scaevola bushes. They are truly creatures of evil.

They carry in their feathers as parasites creatures nearly as malevolent in appearance as themselves—louse flies which look like giant, flattened black house flies. When these are shaken off they sometimes fly to small black automobiles which they mistake for their hosts.

The nests of the frigate birds are coarse, soil-cemented affairs constructed haphazardly of twigs and driftwood. During showers, the cement of this filthy building material dissolves away, allowing eggs to fall to the ground. Nesting material evidently is rare and highly prized, giving rise to theft. A bird in flight occasionally filches a loose piece from a carelessly guarded nest. The iwa will stoop to murder and cannibalism, flying off with an egg or newly hatched young to eat on the wing. There usually is one egg to a nest, entirely white and a little larger than a chicken egg.

Both sexes take turns sitting on the egg and later brooding the growing chicks. This is necessary not only to incubate the egg and keep the chick warm in cool weather, but also as protection against too intense sunshine. At the incubation time the males are resplendent with blood red, semi-transparent throat pouches blown out like balloons. These extend forward to the beak and downward to hide the breast. The color is due to innumerable blood-filled capillaries in the tissues of the pouch.

Not far from the rookeries of the iwas are those of the stupid, red-footed boobies, or gannets. The name booby is from the Spanish word "bobo", meaning "idiot". At times the rookeries of the aggressive marauders and the boob-victims overlap at the edges.

The frigate birds, according to a report of the Pacific Science Board, "escort the stupid, spoon-billed gannets out to feed on schools of squid and small fish. When the gannets get craws full and set sail for home to feed their young, the cruel, curve-billed iwas dive screaming after them, seize them by the tails, and sling the food out of the mouths of the smaller birds. This the iwas scoop up on the wing. This goes on from dawn to dusk. The war cries of the frigates and the plaintive screams of the fleeing gannets quiver down the trade winds like the wailings of lost souls."

It is commonly reported that frigate birds, lacking webbed feet, never land on the surface of the water because they cannot take off again. This is not true; small flocks are frequently seen landing playfully on the Canton island lagoon, floating, and rising again seemingly without any effort whatsoever.

"The birds nesting in the scaevola," says the report, "are tame or, depending on the point of view, too innocent or stupid to fly from their nests when approached. The explanation for this habit is their nesting from time immemorial in areas where no predatory animals, two or four legged, ever have existed. (This, by the way, is a notable characteristic of bird life in the Antarctic. The notorious skuas, with whom even the frigates could hardly compare for blood-thirstiness, will not even bother to move when men pass through a flock of them on the ice.) Tame birds were not killed off but lived to reproduce their kind. Now, unfortunately, Pacific islanders employed as laborers, occasionally club the nesting birds at night preparatory to a feast. Such vandalism and resulting pandemonium in the rookeries should be stopped by legislation."

The ancestors of these and other kinds of sea birds have inhabited the islands during the nesting seasons for milleniums, catching fish and other sea life as food for themselves and their nestlings.

EAGLES AS INDIAN PETS

The proud eagle was once kept as a "domestic animal." Memories of this practice have been obtained from the Shoshoni Indians of the Nevada desert. As recently as fifty years ago individual Indians owned eagle aeries in the mountains. These constituted about the only private property recognized by the tribe and rights were zealously maintained.

Expert climbers who scaled the cliffs took the young eagles from their nests. They were subsequently reared in cages or tied to rocks. The purpose was to harvest their feathers for arrows, decoration, or magical rites. The birds were fed pocket gophers and young groundhogs.

When the birds were full grown the feathers were plucked. Then the captives were taken to the top of a cliff and released.

THE GIANT INSECTS OF THE CAROLINES

Giant walking sticks seven to nine inches long, titan spiders that walk on water, little black crickets that dive and swim long distances under water are some of nature's curiosities on mountainous, jungle-covered Kusaie, easternmost of the Caroline Islands.

Especially unusual are the winged-blue-and-green walking sticks with their fantastic hand-over-hand way of walking. Among the largest of all insects is a walking stick found on the nearby island of Truk. It is reddish-brown and wingless with a body nine inches long. The huge spider's usual abode is the foliage of long grasses overhanging jungle streams. There it lies in wait for the insects which are its usual prey. When alarmed the big spider drops off the grass into the water and starts running swiftly over the surface. It is provided with "water shoes," bristle arrangements on its feet. Probably it does not even get its feet wet.

The submarine crickets are little black insects about an inch long which live on damp basalt rocks along the sides of, and in, the streams. They are almost invisible in the dim jungle light but make themselves known by their continuous chirping. When frightened they make long, high dives from the rocks and swim for undetermined distances a few inches under water, where they are invisible.

By far the most fantastic spectacle found on Kusaie is that of the ghostly light which marks the banks of rivers. It is due to some species of ground-growing fungus. A Smithsonian party once was overtaken by darkness high in the mountains where no trails could be followed through the dank jungle. They started wading down a stream which, they knew, eventually must lead to the lowlands and the coast. They waded, sometimes neck deep, in a tunnel of overhanging branches through whose thick foliage no light could penetrate. But always, glowing on both sides of them, were the lines of luminous fungi.

THE VALLEY WHERE DUSK IS DEATH

A belt of poison night where death strikes with the dusk extends down the western slope of the Peruvian Andes. This death belt, first reported by a Spanish physician in 1630, consists of a few narrow valleys at an elevation of from 3,000 to 8,000 feet in an arid, very desolate and sparsely inhabited country. Nearly everyone who spends a night there is afflicted a few days later by a severe anemia which often proves fatal. This is the "verruga" disease. The red blood cell count drops very rapidly. It is not known whether the cells actually are destroyed by the disease, or whether it inhibits the forming of new ones from the bone marrow. The effect in either case is the same. The blood loses its capacity to carry oxygen and the victim slowly smothers.

The malady is known as Carrión's disease. In 1885 a Peruvian medical student named Carrión inoculated himself with it to prove its identity. He succeeded in showing the cause, at the cost of his own life. He had been inspired to the foolhardy act by extreme patriotism. The Chile-Peru war was just over. Most work on the disease had been done by Chileans. Carión desired that the credit for medical research should come back to Peru.

If one recovers from the anemia a second stage of the malady sets in. The body is covered with wart-like growths, presumably due to some alteration in the blood supply to the skin. One attack gives immunity for life, but the death rate during the first stage is very high.

During daylight the death belt is perfectly safe. This has long been recognized by natives who travel through it freely between sunrise and sunset. The only permanent inhabitants of the region are persons who have recovered from the disease. The borders are sharply defined within a few yards of altitude.

For some years it has been recognized that the infection comes from the bite of a single species of sand fly—a vicious pest smaller than a mosquito. Protection is afforded only by special screens. Ordinary mosquito netting is worthless. The death belt is a place of bright sunshine nearly every day. The insects cannot endure light. They remain secluded and it is difficult to secure specimens, even when the hiding places are known. As soon as darkness comes they emerge in enormous numbers.

Harvard entomologists who investigated the death belt a few years ago spent the hours between sunset and sunrise in a specially screened railroad car. A few moments outside might have proved fatal.

Due to some delicate balance of nature this sand fly seems to be confined almost exclusively to this locality. It is credited with causing about 7,000 deaths in the decade before the last war.

ENIGMA OF EVOLUTION: THE SNAKE

Snakes once had legs. There is evidence in their anatomy that they are descended from four-legged land animals. This evidence is found especially in certain bones near the base of the tail of one of the largest of living snakes, the python, which is the most primitive of the order and presumably nearest to the hypothetical ancestor.

Although the snake remains an enigma of evolution, there is no doubt that it got rid of its legs because they were a distinct hindrance to its peculiar ways of life.

The serpent is not very ancient, as animal types go. Evidently it first appeared in the Cretaceous geological period, about 100,000,000 years ago, when the great dinosaurs were the earth's dominant animals. There are, however, no unquestioned fossils of snakes from the dinosaur days. The first snake-like creature known is represented by fossils from the Eocene, or "dawn", age in North America. This was quite lizard-like in bone structure. It lived about sixty million years ago, when mammals were developing on earth. Rocks in Germany, laid down about twenty million years later, yield fossils of true snakes of the generalized viper type. Sometime later come fossils of snake giants from Egypt. Some of these probably were sixty feet long. But all these were real snakes, with no traces of external limbs. The ancestor seems lost forever because snake skeletons are brittle and delicate and do not easily fossilize.

Having discarded legs, serpents evolved means of locomotion suitable to their ways of life. This has sometimes been described as "walking on the ribs." It requires a highly intricate coordination of ribs and muscles and can be compared best to rowing a boat.

"The life of a serpent," according to Dr. Alfred Leutscher of the British Museum of Natural History, "is a matter of adjustments for what it has lost. It cannot masticate its food so it swallows it whole. It can put a healthy human appetite to shame yet it can, if forced to do so, starve for more than a year. Limbs are missing, so it walks on its ribs, swims and grips with its tail, and climbs with its scales. The outer skin does not grow, so from time to time it is peeled off neatly, even to the scales over the eyes. Taste is poor, but this is compensated for by a strong sense of smell, in which the harmless tongue assists by catching the smell particles from the air. It is proverbially deaf, but may receive ample warning of danger from vibrations through solid objects, which reach its sensitive skin more swiftly than sound can travel through air."

THE FASTEST GROWTH ON EARTH

In the beginning was vestureless life. It was the capacity for self perpetuation and growth in nature, the property of a single complex chemical mixture—protoplasm.

This protoplasm may have come here from another star, a single grain of cosmic dust blown out of the infinite. It may have been mixed by chance in the warm seas of the earth at the beginning of time. It may have been put together according to the design of some cosmic intelligence. It tended to segregate into billions of trillions of infinitesimally minute particles, each sufficient unto itself. The particles were purposeless, voracious, irresistible and immortal. They threatened to devour space and time and all that was in them.

A cell culture of elemental, inchoate life stuff whose original substance increased theoretically 10,000,000,000,000,000,000-fold in forty weeks has been described by Dr. Phillip R. White of the Rockefeller Institute. In his experiments he started with a pellet about the size of a grain of mustard seed cut from a wart-like excrescence on a tobacco plant. He watched it multiply until, arithmetically speaking, if no part had been discarded it would have been an unorganized, purposeless monster spheroid of life 600,000,000 miles in diameter, comparable in size to the whole solar system inside the orbit of Pluto.

It had twelve weeks to complete its first year. At the same rate of growth it then would have been a lusty infant the size of 400,000 solar systems. In a few more weeks it could have swallowed the whole Milky Way galaxy. By the end of its second year it would have filled all the space in known creation, consumed the substance of all the galaxies, and perished of starvation as it bulged outward into the emptiness of infinity.

Such a nightmare actually happened, in reverse. Dr. White had to do everything in a few test tubes, but he was able to witness such a phenomenon of growth as man had not hitherto imagined. First he placed his pellet in a special nutrient solution. It began to expand by the continuous process of splitting in two. Two cells become four, four eight, and so on infinitely. After about two weeks Dr. White cut away a few pellets from the original mass and discarded the rest. These were placed in new nutrient solutions. Every two weeks the experimenter would discard the bulk of each mass which had accumulated and start new cultures with the few pellets which he saved. Each culture increased in size about fifty percent a day. At the end of forty weeks he was left with something not much bigger than he had at the start, but the actual original pellet constituted only about a ten-quintillionth of the final mass.

He happened to have found in the tobacco excrescences an undifferentiated kind of life. The cells had no specialized function. In the actual plant they were kept in order by the rest of the plant cell community, which has no use for cells with no job to do. Once in the nutrient solution, however, they were free of all inhibiting influences. They were not, and never became, wood cells, bark cells, pith cells, leaf cells or any of the other numerous, specialized kinds of cells which make up the plant world. They were something very close to the primaeval plant cells from which, in the course of a couple of billion years, all the others have been derived. Very early these unit structures of life learned that they must stick together and do specialized jobs for each other under the actual conditions of nature. Out of these combinations of specialists has arisen all the magnificent structure of the living world.

But the experimental cells at the Rockefeller Institution had nothing to do except eat and multiply. Each of them was potentially immortal. It did not die but renewed its youth when it had reached its growth by becoming two baby cells. That is how life might have developed from the beginning except for the fact that a cell must eat to live and ordinarily does not have any accommodating scientist to feed it.

BIRDS THAT DUEL

Birds that hold fencing tournaments are the big-billed toucans of Barro Colorado Island, the Smithsonian Institution's tropical preserve in Gatun Lake, Panama Canal Zone.

They fence with their formidable beaks but seem careful not to hurt one another. One scientist who studied Barro Colorado's bird life described the birds as follows: "I saw fourteen toucans scattered about in a big leafless tree in the center of the jungle. Two appeared to be fencing. They stood in one spot and fenced with their bills for a half minute or so, rested, and were at it again. Presently they flew off into the forest and then I noticed two others that had now begun to fence. Then one of these flew away, and the remaining one picked a new opponent and fell to fencing again....They did not move about much while fencing, although sometimes one climbed above the other as though to gain an advantage. They fenced against each other's beaks and never seemed to strike at the body. There was a fairly rapid give and take...the bills clattering loudly against each other."

These fencing toucans are among the more conspicuous birds of the island, particularly because of their call—a shrill, froglike "cree," which is repeated over and over again and can be heard half a mile away. The call is most frequent in the morning and late in the afternoon, but it stops abruptly at sunset.

BRAKES ON PLANT LIFE

There is a "brake" on plant development—perhaps one of nature's most fundamental controls over surging life. It is a relatively narrow band of light on the edge of the invisible infrared in the solar spectrum. Plant life, and through plants all life, is tied intimately to certain solar wave bands. It has long been recognized that the cornerstone of all life on earth is the process of photosynthesis by which plants, through energy provided by sunlight, are able to synthesize carbohydrates from water and carbon dioxide taken from the air. Animals eat these carbohydrates, the basic food. Other animals eat the carbohydrate eaters, and thus the chain extends from the simplest organisms to man.

But without some other process the carbohydrates might be a formless mass. The second process is that which shapes a plant and controls development of stems, leaves, and blossoms. This may be a light effect second in importance only to photosynthesis itself. It requires very little solar energy. Smithsonian Institution experiments have demonstrated that the control is exercised by red light with a maximum of efficiency at wavelengths around 660 millimicrons—or millionths of millimeters. It has been demonstrated, however, that this formative action can be blocked effectively by irradiation with wavelengths in the far red. The greatest effect is at wavelengths between 710 and 730 millimicrons.

The "brake" is not applied immediately. The maximum efficiency of the far red energy occurs a little more than an hour after the plant is exposed to the formative wavelengths. The implication is that the action interferes with the development process by acting on some product the formation of which is initiated by the shorter red wavelengths. The experiments have been carried out with seedlings of beans. In other experiments it has been found that damage to plants from X-ray exposure—insofar as this results in breaking the bundles of genes, or units of heredity—can be increased from 30 to 50 percent by previous exposure to about the same wave band of far red light that reverses the formative process. On the other hand, the increase in damage is nullified if the X-ray exposure is followed by exposure to the red wave band.

Breaking of the chromosomes, or strings of genes, is one of the first evidences of damage to living organisms by exposure to ionizing radiation. This breaking is responsible for some of the adverse hereditary effects concerning which there has been a great deal of discussion because of possible effects of the atomic bomb fall-out.

The experiments were carried out with pollen of flowers and root tips of beans where results are relatively easy to determine.

SNAILS ARE THE FLOWERS OF THE SEA

There are more than 80,000 kinds of snails in the world. They swim, jump, crawl, burrow, live at the bottom of the sea and in the tops of trees. They range in size from the horse conch of Florida, two feet long, to animals hardly the size of a grain of sugar. About half of all species live in the seas.

Most are bottom dwellers, unable to swim, but several spend their lives on the surface. One, the purple janthina, floats upside down on a raft of air bubbles trapped in a special kind of mucous which it secretes. Others live permanently attached to sea weeds. Most abundant of the sea snails probably are the pterepods, or sea butterflies, which live several feet below the surface in daylight but come to the top in countless hordes at night. In some places the sea bottom is littered many feet deep with their shells, of which there is almost constant rain as the animals die.

Loveliest flowers of the sea are the nudibranchs. Seldom has nature produced in either plants or animals such elaborate combinations of brilliant colors and decorative appendages as in the bodies of these shell-less ocean snails. Although there are more than 2,000 species, they are among the least known of all sea creatures. One reason for this is that most of them are quite small, ranging from a fourth to half an inch in length. Their coloring hardly can be appreciated except under some magnification.

Nowhere are they very abundant. Their habitats vary from close inshore to deep water, but they are most likely to be seen in pools left among shore rocks by receding tides. Their extremely elaborate color patterns may be protective, to some extent. It is known that certain species have the ability to change colors in response to changes in their environment. They become bright red, for example, when living in association with a red sponge. Even more decorative than the color patterns are the appendages, extensions of the skin and sometimes of the digestive tract, which take the forms of delicately modelled, almost microscopic plants.

All these nudibranchs are flesh-eating creatures feeding chiefly on sea anemones found on the sea bottom. Most of the anemones are equipped with thousands of so-called nematocysts or stinging organs. These are microscopic, ball-shaped structures filled with a virulent poison. The same mechanism is best known in sea nettles. As soon as a nematocyst is exposed to any tension it explodes, releasing this poison.

The little sea snails have evolved the ability to swallow the poison balls without exploding them. They pass into the digestive tract, but are not digested. In some way the nematocysts find their way through certain of the appendages growing out of the digestive organs to the outside of the body. There they are retained, and serve the sea snail in the same way they served the sea anemone. The little creature becomes quite dangerous to any of its natural enemies.

Among the most enthusiastic nudibranch collectors is the Emperor of Japan, who has discovered and described several new species. Some of his publications about them have been illustrated by leading Japanese artists and show the unearthly beauty of the creatures to the best advantage.

THE BRUTAL SOUTH POLE BIRDS

The southernmost birds on earth—the only higher animal except man and his dogs that go close to the South Pole—are the Antarctic skuas. They are fierce, brutal little killers. Dwellers in the earth's most inhospitable habitat, they have been able to survive largely because of their extreme rapaciousness.

All other Antarctic birds, such as the penguins, stay close to the shore of the desolate continent. The skua has been seen at least 300 miles inland, and occasionally may fly across the pole itself.

These birds arrive on the coast of Antarctica about the middle of October, the beginning of the southern summer, after spending the winter north of the circle. Their arrival is timed to coincide with the egg-laying of the Adelie penguins. The skua's chief food consists of penguin eggs and chicks which it devours by the hundreds. Scores of half-eaten, trampled bodies of young penguins always can be found during the hatching season near the sites of penguin rookeries. The skua is hardly a match for the parent birds but is expert in separating chicks from the brood and killing them when they have no protection. It is a creature of relatively enormous strength and endurance and flies long distances carrying chunks of meat bigger than itself. It also is an extremely noisy, quarrelsome creature—an outstanding example of the philosophy of every individual for itself. There is no brooding of chicks nor protecting them from the elements. The parents hardly bother to feed them.

Little skuas, it is reported, come out of the eggs fighting. Usually there are two eggs to a nest. One chick probably is a trifle weaker than the other. In a short time it is driven from the nest, killed and eaten by its rapacious brother or sister. It may even become the prey of its own hungry parents. Skuas also have the habit of eating their own eggs. This keeps the population within the limits of the food supply.

SILK-BEARDED CLAMS

Jason's golden fleece may have been woven from the beard of a silk-bearded clam. The same sort of cloth, in fact, still is produced on a small scale in Italy, chiefly for the tourist trade. A silk glove of modern manufacture now is in the Smithsonian collections.

The clam is a giant Mediterranean species, the pinna marina. Its shell reaches a maximum length of about three feet, but the average is less than half this. From a gland in its "foot" it secretes milk-like strands with which it attaches itself to the sea bottom. These strands are as much as a foot long.

The silk is of exceptionally fine quality—at least it was so regarded by the Arabs who maintained centers for manufacture of the cloth in Spain, Italy and North Africa. Says one Arab author: "At a certain time of the year an animal comes forth from the sea and rubs itself on the stones of the seashore. A down soft as silk with a golden color falls off it. It is fine and small and garments are woven from it which take on different colors during the day. The Umayyad kings (of Spain) used to put restrictions upon it so that it was only exported secretly. The price of a garment is more than 100 dinars, on account of its fineness and beauty."

The value of a dinar—the gold coin of the Moslem world—is difficult to calculate in any present coinage, but it was at least the equivalent of a dollar.

Says another Arab writer: "I have seen how it is gathered. Divers dive into the sea and bring out tubers like onions with a kind of neck which has hairs on the upper part. The tubers like onions burst and let forth hairs which are combed and become like wool. They spin it and make a woof of it so as to pass a warp of silk through it. The most magnificent royal garments of Tunis are made of it."

Gigantic clams, nearly five feet long and weighing more than 400 pounds, who raise crops of microscopic plants for their own sustenance are among nature's fantasies found on Australia's Great Barrier Reef. These molluscan titans have formed a curious partnership with the zooxanthellae, a family of microscopic algae. The plants live as parasites in the blood cells of the inner lobe of the clam's mantle. Upon this mantle is a lens-like structure which looks like an eye. These mollusks, however, are blind as any other clams and the eye-like protuberances, it has been determined, are only windows by which light is admitted to the parasitic algae within the blood cells.

The surplus of algae is carried by the blood stream to the clam's digestive organs where it serves as food.

Another giant clam, the tridacna of East Indian seas, may weigh up to 600 pounds. The monsters constitute a peril for divers who unwittingly step inside the open valves. These snap shut, imprisoning the diver's foot and, unless he can get help, he is held in the trap and drowned.

PEARLS GROW IN BROOKS

Excellent pearls occur occasionally in fresh water clams. A pearl of perfect form and pure color was found in such a clam taken from a brook near Paterson, New Jersey, in 1857. It sold at Tiffany's for $1,000 and shortly afterwards was resold in Paris for $2,200. This started pearl hunts in brooks all over the country.

On the arrival of Europeans in Florida, Louisiana and Virginia, fabulous legends were circulated about the enormous treasures to be obtained by plundering Indian graves. A contemporary chronicler of the audacious DeSoto expedition, reported that the conquistadore got 350 pounds of fine pearls at the Creek town of Cofitachique on the Savannah River.

A member of the first Virginia colony "gathered together from among the savage people about five thousande; of which number he chose so many as mayd a fayre chain; which for their likenesse and uniformitie in roundnesse, orietnesse and pidenesse, of many excellente colours with equalities in greatnesse were verie fayre and rare."

The supply, however, was quite limited. Indian pearls were the subject of a special study by the late Dr. William H. Holmes. "The majority of those obtained," he reported, "were ruined as jewels by the heat employed in opening the shellfish from which they were abstracted. Many of the larger specimens probably were not real pearls but polished beads cut from the nacre of sea shells and quite worthless as gems. It has been found that the real pearls were obtained from bivalve shells—from the oyster along the sea shore and in tidewater inlets and from the mussel on the shores of lakes and rivers.

"But the very general use of pearls by the pre-Columbian natives is amply attested. More than 60,000, nearly two pecks, were obtained, drilled and undrilled, from a single burial mound near Madisonville, Ohio."

GRASSHOPPER-INFESTED GLACIERS

Among America's natural curiosities are "grasshopper glaciers." These are great masses of glacial ice containing layers of imbedded, frozen grasshoppers. Such layers are probably remnants of vast migrations which have taken place at irregular intervals over several centuries. Great hordes of the insects either flew over the glacier or were carried there by winds, and while there sudden snow storms or cold air rising from the ice field caused them to drop. They were imbedded so quickly in the falling snow, which later became ice, that they have remained perfectly preserved for centuries. The most notable of these glaciers is in the Beartooth mountains of Montana. Others have been reported from the high mountains of Africa.

MONSTER CLAMS OF POLYNESIA

Largest of clams and largest of all shellfish is a native of Polynesian seas. The two halves may weigh as much as 500 pounds. The flesh is eaten raw by natives. The interior of the shell is like polished marble. Such shells frequently were used as founts for holy water in European churches. A particularly large one attracted much attention in the Church of St. Sulpice in Paris. Such clams are found at depths up to 17 fathoms. They fasten themselves to rocks by a process so tough that it can only be severed with an axe.

CORALS COMBINE PLANTS AND ANIMAL LIFE

A coral reef is a gigantic "plant-animal." It is a community of countless billions of plants and countless billions of animals which act as a single organism, like the countless millions of specialized cells that make up the body of a man or a mouse. It is probably the most efficient of all earthly creatures. It is self-sufficient, creating its own constant food supply. It is essentially immortal. It is hungry like an animal. It is motionless like a plant. It is both and combines the attributes of both. It is the largest and most enduring of all creatures of land or sea.

The animals are coral polyps. They are tiny, wormlike organisms with mouths surrounded by constantly probing tentacles. They are rapacious and insatiable. They are essentially voraciously hungry stomachs, bloodless, brainless, sightless, heartless. The polyps are close to the bottom of animal life, vaguely related to the white, stinging sea nettles which are the scourges of summer beaches. These little creatures extract lime from sea water and secrete for themselves limestone "houses," the "bones" of the superorganism. Out of these they have built up islands and almost subcontinents. Sharing their limestone cells are quite unrelated organisms, single-celled plants or algae. These plants possess the green of grass and forests, whose molecules create out of carbon dioxide and water through the energy of captured sunlight starches and sugars which are the fuel of animal life. This process of photosynthesis is the cornerstone of all life on earth.

Thus the plants feed their partner animals. The excretion of the animals, in turn, provides the essential fertilizer of the plants. Considering the coral reef as a superorganism one might almost say that it eats itself but loses nothing in the process. A reef, considered as a superorganism, represents about the last word in nature's efficiency. It has been found, for example, that one acre of coral reef produces about 74,000 pounds of sugar a year, a record barely reached by man on his most efficiently managed plantations. All this sugar is devoured by the polyps. Apparently the fertility of the surrounding sea makes little difference. Coral reefs flourish in parts of the ocean that are essentially deserts.

A marine biological laboratory has been established by the U. S. Atomic Energy Commission, to study effects of the radiation from nuclear explosions on plant-animal populations. The first requirement has been to determine the natural condition of the organisms before being subjected to this radiation. Then whatever changes take place with subsequent bomb tests can be noted. The work has been undertaken by biologists of Duke University and the University of Georgia. Such a life community, both a vast assembly of organisms and a sort of superorganism, is an almost perfect subject for the required observations. The first job, according to the commission report, has been to measure the "basal metabolism" of the reef as a whole.

Admittedly the conception of a reef as a sort of superorganism is somewhat mystical. The Duke and University of Georgia biologists do not maintain that there is any consciousness of constituting a whole on the part of the individual organisms. It is likely that they have no consciousness of anything. The outstanding fact is that they behave so much like a whole.

A reef is an outstanding example of the two major divisions of life, plant and animal, working in perfect co-operation. The actual co-operation of plant and animal in an integrated organism is not unique for the coral reefs. Something of the sort occurs in certain sea worms, near the bottom of the worm family, that grow green algae in their blood streams. These worms make some of the beaches of Normandy grass-green in summer. The algae are necessary for their existence. There may be a few other examples throughout the animal kingdom.

THE FIRST ENGINEERS—TERMITES

Termite civilization probably has reached its greatest heights in architecture and engineering. Australian mounds, built by workers out of earth particles cemented together by a salivary gland secretion, are steeple-shaped, as much as twenty feet high, and with bases twelve feet in diameter. Hundreds of such structures may be scattered over a few acres. Such an assemblage looks like a large native village, although architecturally the structures are far beyond the abilities of primitive man. The common type consists of a solid, hard outer wall which has the strength of superfine concrete. It is almost impossible to break through this material. Immediately inside are numerous thin-walled passages and galleries. Below these, at the ground level and about in the center, are the quarters of king and queen and the nursery. From the mound, passages for the food foragers lead in all directions through the soil. A mound two feet high will house approximately two million individuals.

Long before architects, termites developed the art of air conditioning. Proper humidity inside the nest is essential to the existence of the soft-bodied workers. The majority of species, however, are found in latitudes with long, dry seasons. To meet such conditions the insects achieved humidity control in various ways still not understood. Notable are the structures of the Australian compass termites who erect dwellings eight to twelve feet high with flattened sides. The broad ends always point east and west, the narrow ends north and south. These nests are strong enough to support the stamping of wild bulls. A group of them looks like a particularly well-constructed native village, or the site of some extinct human civilization. Apparently the precise orientation of the nests is associated with prevailing winds and in some way contributes to maintaining a constant humidity.

The blind creatures seem to have developed special sense organs, unknown to man and probably unique in the animal kingdom. One of these is reportedly a brain barometer which is extremely sensitive to slight humidity changes. Both soldiers and workers respond with military precision to any threat to their neighbors. This believed due to an extreme sensitivity to vibration.

Few varieties of termites can endure sunshine. Some construct paperlike umbrellas which they carry with them when they come above ground. One species on Barro Colorado island in the Panama Canal Zone which attacks live trees first builds a thin earth crust around the trunk, seven to eight feet from the earth. Beneath this crust they seek out weak spots in wood which enable them to penetrate into the heart of the tree.

Termite armies, in distinction from those of ants, serve only as defensive forces. There are two kinds of soldiers. Some are equipped with enormous jaws with which to rend the enemy. These are so tenacious that when the body is torn away from the prey the mandibles remain in place. Others are the bayonet men and chemical warfare troops. These fighters have a protrusion on the front of the head which looks like a long nose but which actually has developed from a primitive eye.

From this protrusion a sticky acid is exuded. In rare instance it may be spurted a short distance—an inch or less. These soldiers fight battles to the death with war-like ants which invade their nests. The termite warrior rams with his nose-like organ the so-called "pedicle" of the ant, the narrowest part of its body, smearing it with the liquid. This never has been completely analyzed. It is a powerful acid, but is not the well-known formic acid exuded by ants. It has strong corrosive properties when applied to metals. It has a pungent odor which, however, is characteristic of all termites and the ancestral cockroaches.

Between ants and termites there is perpetual war. Army ants, especially, try to raid termite nests to feed on the young whenever they can find any crack in the walls through which they can squeeze their bodies. But when there is any break in the nest the termite soldiers immediately arrange themselves in a circle around the opening while workers bring up little slabs of earth from the interior to patch the wall.

Most common of the Barro Colorado species are the amitermes which build hemisphere-shaped red mounds about two feet in diameter. These are made of tiny particles of earth which have passed through the alimentary tracts of the insects where they are coated with a cement-like material. Such a nest is impervious to water. It is so sturdy that a heavy man can jump up and down on it without breaking the roof. It cannot be broken open with a machete.

Another common species build the so-called "niggerhead" nests, about the size of footballs, on fence posts and trees,—especially dead trees whose stumps protrude out of Gatun Lake. These nests also are extremely sturdy. They are made of a mixture of earth grains and finely digested wood. From such a nest numerous runways traverse the trunk, sometimes connecting with smaller colonial "niggerheads."

OYSTER ODDITIES

An oyster can change its sex several times during its life. This has been determined by Dr. Paul Galtsoff of the U. S. Fish and Wildlife Service by observing an experimental colony. In the first year 8% of the males changed to females and 13% of the females became males. In the second year 11% of the males changed sex and 12% of the females. One sex change, Dr. Galtsoff found, makes the same individual more likely to undergo another.

A single Pacific coast oyster produces approximately 10,000,000,000 descendants a year. If all survived in five generations they would constitute a mass eight times the size of the world.

Clams and oysters appear to be about the most stupid animals in creation. Actually each has three "brains," or nerve ganglia. One controls the feeding apparatus, another the viscera, and a third the utilization of oxygen.

THE WORLD'S BIGGEST SNEEZE

The sneeze of the elephant has been described as "like the bursting of a boiler of considerable size." When the elephant feels the onset of one of these titanic eruptions it appears to realize that a momentous event is about to take place. It becomes extremely restless and is seemingly unable to stand still for a moment. The sneeze is preceded by a tremendous, wall-shaking bellow.

Although elephants are subject to frequent colds the sneeze is a rare phenomenon. For this reason it is regarded as a good luck sign, especially among Moslems of India, who gather around and wait patiently for the event. When it starts they bow their heads and pray for the realization of their wishes.

THE LUMINESCENT CTENOPHORES

There are windless nights when Caribbean waters seem like fields of green fireflies. This is due to vast numbers of luminescent ctenophores or comb-bearers. One the most abundant and least known forms of animal life, they are also among the most delicate. Although they are related to the planarian worm and the jelly fish, they are quite unique.

Superficially they seem little more than animate bags of water with skins thinner than the most delicate tissue paper. They abound in staggering numbers over most of the world. One of the most familiar types is the American mnemiopsis. On calm summer days the amber green species sometimes covers completely thousands of square yards of sea—like a raft formed of millions of individuals floating just below the surface. A classic ground for this phenomenon is Narragansett Bay.

Like the rest of its race, this ctenophore is like a fragment of moonlight on the sea. It is so fragile that the slightest current of water in its neighborhood is sufficient to tear it to bits. It is about as elusive as moonlight. When grasped gently the jelly-like substance slides through the fingers. Taken in a net and placed in salt water it vanishes completely on the way from boat to laboratory. Intact specimens are almost unknown in scientific collections.

Ordinarily they live at considerable depths in the zone of absolute calm where all wave movement ceases. Great hordes rise to be the surface only on nights when the surface of the ocean is like a sheet of glass.

They are among the loveliest of all sea creatures. The delicacy of their coloring is that of spring arbutus or anemone. Their presence is indicated chiefly by the brilliant flashes of rainbow colors as they pass a few inches below the surface.

The majority are pear-shaped. Giant of the race is Venus' girdle, best known in the Mediterranean but found in most sub-tropical seas and sometimes swept as far north as the coast of New England. It is an undulating, iridescent ribbon as much as five feet long and two inches wide. The mnemiopsis of southern New England waters is ball-shaped with a diameter of about four inches.

Ctenophores are most varied in the Bay of Naples; there 18 species have been identified. There are 14 species now known in the Caribbean. In absolute numbers, however, the fragile creatures are most abundant in North Atlantic and sub-Arctic waters where, because of ordinarily rough seas, they seldom are seen. There they constitute one of the major menaces of the cod fisheries. Despite their fragility they are vicious little animals, devouring cod eggs and fry in incalculable numbers.

Each living water bag has a slit-like mouth on top and what apparently is a sense organ of some kind on the bottom. The minute, struggling prey are seized in two pincer-like tentacles and pushed into the mouth. They are digested quickly by the juices in the water sack in which float about whatever vital organs the Ctenophore possesses.

The ctenophores are by no means aberrant jellyfish, which they resemble only in the extreme tenuousness of their bodies. They have no umbrellas and no stinging cells. Two forms are known which have flattened bodies like planarian worms and which creep on the sea floor. Because of various similarities in the development of both creatures some zoologists believe they are immediate descendants of a unknown common ancestor.

The function of their weird green luminescence is unknown. It would seem of questionable value in attracting prey and it is difficult to imagine that these most fragile and evanescent of earth's creatures have any sort of love life. Nevertheless lightmaking seems to constitute a purposeful part of their activities.

THE FOREST THAT TIME FORGOT

Knee-high red and pink ferns fill the jungle hollow. Around them are green leaves covered with parallel white lines in sets of five with dots on the lines which look like notes of music. These leaves are known as "music paper." There is no record that anybody has tried to play the tunes nature has written on them.

Mixed with them are "sandpaper leaves" with surfaces so rough that they are used locally for the same purpose as sheets of sandpaper elsewhere. Sinister hangman's ropes swing, as if awaiting their victims, from branches along the jungle paths.

Such are a few random notes from a cloudland jungle—in many ways like a forest of prehistoric days—in Venezuela's Henry Pittier national forest. Here flourishes the giant tree fern, most characteristic tree of the vast ancient forests from which coal deposits were formed. In the tree fern fronds lurk worms and amphibians not vastly different from the tree creatures of the Devonian geological area.

This is a forest of the central tropics. Paradoxically it is also, when seen from a little distance, a New England forest of late September with groves of straight, white-trunked palm trees which look like birches and patches of flame color in the treetops which look like maple leaves starting to put on their autumn coloring. The temperature, in fact, is about that of a warm Autumn day in New England, especially as dusk comes and a white veil of mist rolls over the mountaintops from the sea.

The patches of flame color which look like maple leaves are orange and red blossoms of the gallito or "cock flowers," so called because the bloom resembles so much the body of a miniature rooster. The gallito appears high in the treetops. It is about the most abundant and conspicuous flower of the cloud jungle. It grows on big, grey-trunked trees whose bark looks like rough-woven linen. Each blooming tree is filled with brilliantly colored humming birds and red and green parrots.

Trees in the high jungle hills wear thick green overcoats of moss and lichens. There is one dark-green form of moss which grows about an inch high and looks like a miniature cedar leaf. Many of the older trees, especially palms, are "rusty" with a species of red lichen which spreads rapidly over the trunks. Among them is a blossoming tree with a straight, spined grey trunk from 30 to 40 feet high which is a close relative of the potato.

The cloud forest is predominantly the home of the epiphytes, such as long, dangling masses of red, pink and pearl orchids which grow on the trees. They require plenty of moisture. In this mountain swamp the trees always are soaking wet. This is an ideal environment for the eight or ten varieties of moss which grow so luxuriantly.

There are green-walled cave openings ten feet high and ten feet wide in the bottoms of the trunks of giant trees. Exposed roots lie across the paths, covered with moss in which there are leprous white spots. They look like enormous, writhing malevolent green serpents.

THE VERSATILITY OF THE ELEPHANT'S TRUNK

The elephant's trunk is a tool surpassed in effectiveness only by the hand of man. It is a muscular prolongation of combined nose and upper lip, which have grown together. It is associated closely with the motor and sensory centers in the brain cortex and is under such delicate voluntary control that with its enormous strength is combined extreme fineness of movement. The trunk terminates in one of two fingerlike projections which seem capable of almost as delicate voluntary movements as are human fingers.

The trunk is a supernose. As a sensory organ it is the elephant's chief means of securing information about his environment. With it the animal can detect the direction, and perhaps the distance, of olfactory stimuli from all sorts of sources. It is as vital in an elephantine scheme of things as are eyes to a human being.

The trunk is the elephant's chief servant Without it the monster is the equivalent of a blind man. It has approximately 40,000 muscles and a highly developed sensory and motor nerve supply. The organ has enormous strength, sufficient to tear up a tree by its roots.

Here are some of the things the animal is credited with being able to do with the trunk: pick up a pin from the ground, select and secure a single tussock of appetizing herbage, uncork a wine bottle, untie a slip knot, unbolt a gate, throw up and catch a baseball, pull the trigger to fire a gun, ring a bell.

A female elephant owned by the Duke of Devonshire in the 1880's was allowed almost a free range over the park of his estate. She made herself useful by sweeping the paths with a broom and by carrying a garden watering pot. Her most celebrated achievement was that of opening a tightly corked wine bottle. She would hold it against the ground at about a 45 degree angle with one of her front feet and gradually twist out the cork—barely protruding above the neck of the bottle—with her trunk. After emptying the contents into her mouth she would hand the empty bottle to her keeper.

FIENDISH VAMPIRES OF THE NIGHT

About the middle of the eighteenth century belief in vampirism spread like an epidemic across France and England. Dead men hellishly condemned to live forever came out of their sepulchres at midnight, took the forms of various animals, and feasted on the blood of the living (who, in turn, died and became vampires). This was a superstition which previously had been confined largely to Slavic countries. Its influence in France and England seems to have started with tales brought back from the New World by Spanish explorers of actual vampires—sinister, black-winged, fiend-faced flying mammals who actually fed on the blood of sleeping humans. Thenceforth the popular conception of a vampire was that of a large bat, hovering over the unsuspecting, eternally doomed sleeper.

The stories doubtless were greatly ornamented and exaggerated. However, the vampire bat of the American tropics is a gruesome reality. It is now known to be a carrier of the rabies virus.

It is a small, brown bat condemned by nature to live exclusively on blood. Its throat is too small to swallow solid particles. Its stomach is especially adapted for rapid digestion. It feasts on all sorts of mammals, including man, and the incisions of its razor-sharp teeth are so nearly painless that a sleeper seldom is awakened. Supposedly it always bites man on the bottoms of the toes.

The loathsome little creature does not actually suck blood, as long was supposed. Instead, according to observers, it laps up blood with its tongue. Its saliva is believed to contain an anti-coagulant which keeps a wound bleeding for hours. From 20 to 25 minutes is required for a meal, during which the animal gorges itself until its body becomes spherical.

"We slept so soundly", records an Amazon explorer, "it was not until morning we discovered that we had been raided during the night by vampire bats and the whole party was covered with blood stains from the many bites. It may seem unreasonable to the uninitiated that we could have been thus bitten and not disturbed in our sleep but the fact remains that there is no pain produced at the time of the bite, nor for several hours afterwards."

It feeds only at night Like most New World tropical bats, it sleeps during the day in the total darkness of caverns where it hangs in clusters from the ceilings. Such a bat cave, about as gruesome a place as could be found on earth, was explored a few years ago by Dr. Raymond L. Ditmars of the American Museum of Natural History. This cave, which the bats shared with scorpions who had wing spreads of five inches, was found in the Chagras Valley of Panama.

The mammal has a strikingly spider-like appearance. Probably alone among bats it can walk as a quadruped, using its wings as front feet. That, of course, is what they were originally before the grotesque creatures invaded the air.

REMARKABLE ORCHIDS

A flower that opens only in moonlight is one of Venezuela's plant curiosities. It is an ivory white, velvety orchid with a dazzling blossom. For full fertilization it depends entirely on nocturnal butterflies which sip nectar while pollenization takes place.

This curious flower is one of approximately 800 orchid species, some of them among the most beautiful in the world, which grow in Venezuela. Among these is probably the prettiest and rarest of all orchids, the mother-of-pearl flower which can be found, and then only rarely, in the Gran Sabana country at altitudes of more than 3,000 feet. Only a few specimens ever have been brought out by collectors.

Another high mountain variety has square petals with fringed edges. Found in the jungles of the upper Orinoco is an orchid with blossoms measuring up to 16 inches in diameter. A completely unique orchid has been found growing in water. (All other species live as parasites on trees or rocks—or in the soil like other plants.)

Throughout the world there are more than 20,000 species of orchids, the great majority of which are found only in the mountainous regions of the tropics. A few, however, can be found growing as far north as the Arctic Circle.

NATURE'S INSECTICIDE: THE MILLIPEDE

Far leas malevolent than the centipede—and probably a somewhat more primitive form of animal life—is the millipede or "thousand legs". It is a strictly vegetarian creature that lives under stones, logs or in rotting tree trunks and feeds on soft roots, leaves and fruits.

Millipedes are seldom seen. They shun light, although in the tropics they sometimes come out of their retreats after heavy rains and crawl over the ground. The animal has twenty to forty legs, two pair on each segment of the body—a characteristic in which it differs striking from the centipedes to whom it is only distantly related. Movement is in an almost mathematically straight line, with a series of wave-like undulations in which apparently all the legs on one side of the body move in unison. All millipedes are essentially blind. Their eyes are able only to distinguish light from dark, but as they crawl every inch of their path is explored by their delicately sensitive antennae.

So secretive is their life that relatively little is known of their behavior. The female of one European species burrows in the earth, moistens bits of soil with a sticky fluid from the salivary glands in her mouth, and thus makes tiny bricks. These she builds into the form of a hollow sphere, about the size of a walnut, with a hole in the top through which she lays from 50 to 100 eggs. Others lay their eggs in bunches in the soil and coil around them until they hatch. Mothers may even remain with the young for a few days.

The bite of the millipede, unlike that of the centipede, is not poisonous. But the animal has "stink glands" from which a foul-smelling liquid containing the extremely poisonous prussic acid is exuded. This presumably affords an adequate protection against driver ants and birds, the natural enemies. The secretion is so powerful that a couple of millipedes placed in a can kill insects as effectively as a small dose of potassium cyanide.

One member of the race, spirobolus marginatus, as much as four inches long and with a body made up of fifty-seven segments, is fairly common under logs in the northeastern United States. At certain seasons these creatures become restless, leave the soil and come into houses. They may swarm in basements and on ground floors. They crawl up walls and drop from ceilings. These invasions usually take place in the autumn and presumably are associated with migrations to find winter quarters. In some cottages surrounded by trees as many as seven hundred have been counted in a room in one evening. However embarrassing to hosts, it must be realized that millipedes never bite and that they do no damage to furniture. The only accusation yet made against them refers to one species, the so-called greenhouse millipede, which may cause considerable damage to potted plants.

In emergencies the millipede is able to roll itself in a tight ball like its presumed ancestors, the primaeval trilobites. In one Madagascan species this ball is as big as a golf ball. Some millipedes are less than a twentieth of an inch long.

Gigantic millipedes are known from the tree fern swamps of the Carboniferous geological period when the great coal deposits were formed. They were about a foot long and their bodies were covered with long, sharp spines. This apparently was to make them distasteful to the giant amphibians, remotely related to present day frogs and toads, who were the dominant four-footed animals in the world at the time. Thus the millipede has almost as lengthy a history on earth as the more insect-like cockroach of those same forests of 250,000,000 years age.

BATS HAVE BUILT-IN RADAR

Bats "see" with their ears. Echoes of sounds inaudible to man enable the flying mammals to find their way through the almost absolute darkness of deep cavern or jungle. These creatures might be considered inventors of the Navy's sonar device by which underwater obstacles are located by echoes—or even, in a sense, of radar.

Almost entirely creatures of night and late twilight, bats have small and poorly developed eyes. When one is on the wing it emits an almost constant succession of inaudible "squeaks" at a sound frequency of between 25,000 and 70,000 vibrations a second. The human hearing range reaches only to 30,000. Each squeak, according to measurements by Dr. Donald R. Griffin of Cornell University, lasts about two-hundredths of a second. In ordinary flight over open country it is repeated about ten times a second. By means of the echoes it apparently is possible to detect and avoid any obstacle, even one as small as a strand of silk thread strung across the path, within a distance of ten or twelve feet.

The bat does not hear its own squeaks. Each time one is uttered an ear muscle contracts automatically, thus momentarily shutting off the sound itself so that only the echo can be heard. It is possible that each animal has its individual sound pattern and is guided only by its own echoes. Otherwise, it would seem, there would be complete confusion from the echoes of several hundred bats moving in a flock.

Largest of the bats are northern India's flying foxes. The body is shaped almost precisely like that of a small fox and is covered with fine, dark-brown hair. The wing spread is about three feet. These flying foxes move in flocks of thousands. They are exclusively fruit eaters and forest dwellers. They are the only bats eaten by man. Their flesh is said to resemble chicken.

Insect-eating bats are prisoners of the air. Once on the wing they must remain in flight all night until they return to the dark caves where they sleep all day, suspended head downwards. Flying from dusk to dawn requires an enormous amount of energy for which a lot of food is required. One of these animals probably must eat about a third of its own weight in insects each night. Thus it is a good friend of the farmer and one of the potent factors in keeping the balance of nature.

If a bat lit on the ground or on any solid object it would be very difficult, perhaps impossible, to get it on the wing again. This is accomplished only by falling from its sleeping place.

The hibernation of temperate zone bats appears very close to complete lifelessness and is probably the most deathlike sleep experienced by any mammal. Animals close to a cave entrance have been found completely coated with ice, as moisture has congealed on the fur. Yet when they wake in the spring they appear none the worse for the experience.

CRABS THAT CLIMB TREES

A fantastic race of small, pale hermit crabs are the most numerous and conspicuous animal inhabitants of war-wrecked Pacific islands. The multitudes of these crustaceans may have a considerable role, beneficial and otherwise, in present efforts to cover these white sand wastes with grass and trees.

Of all creatures which start life in the sea, hermit crabs have become best adapted to continual existence on land. Like others of their race they are shell-less and soft-bodied. For protection against enemies and against being dried out by the glaring sun, they live in houses—the abandoned shells of other sea creatures which have been cast ashore. They carry their houses on their backs. When a crab outgrows its shelter it moves to a larger one, changing its dwelling four or five times during a normal lifetime. There is never any housing shortage for those in the small stages of growth. However, the sole refuge for the crab which has reached full size is the "cats-eye," the shell of a marine snail as much as three inches in diameter with an opalescent pink inner lining which glistens like the eye of a cat. Only the hermits which can find such shells survive.

In searching for food the crabs climb the trunks and branches of kou trees which grow all over the Pacific islands. They eat the bark along the upper side of the branches; most trees show long scars which are the results of past injuries.

A common habit, especially of the undersized individuals, is cleverly to tear off and eat only the ovaries and stamens of blossoming plants. "These are certainly not isolated acts," says a Pacific Science Board report, "but ones perfected by practice and perhaps instinct. The crabs probably decimate the flora, feeding particularly on tender seedlings. They largely are responsible for the paucity of different kinds of plants on some islands. The seeds of any new kinds of plants washing to its shores are subject to their inspection and, if palatable, sacrificed to their appetite. The foreign plants now being introduced as seeds and seedlings must not only surmount the drastic condition of drought and salinity but also the hurdle of these voracious animals."

In the spring the females carry their numerous maroon colored eggs attached to their abdomens. When do they return to the ocean to allow these eggs to hatch their free-swimming larvae that resemble so closely the shrimp-like ancestor of all hermit crabs? Where do they throw off the hard, non-expanding shells they have requisitioned as they increase in size, in burrows on land or in the ocean? How, with gills adapted for respiration in water, have they perfected respiration on land? Questions such as these are still unanswered.

THE FEROCIOUS CENTIPEDE

"Natives of Brazil call the centipede the ambua. These creatures of a thousand legs, some of which are more than a foot long, bend as they crawl along and are reckoned very poisonous. In their going it is observable that on each side of their bodies every leg has its motion, one regularly after the other; being numerous, their legs have a kind of undulation and thereby communicate to the body a swifter progression than one would imagine where so many short feet are to take so many short steps that follow one another, rolling on like the waves of the sea."

The eighteenth century British naturalist Charles Owen was not alone in considering the millipedes and centipedes as kinds of snakes; nor in being confused, as naturalists still are, at their curious, complicated way of moving. There had been highly exaggerated reports. The Spaniard Ulloa, Columbus' gold assayer, described some centipedes he saw on the northern coast of South America as a yard long and six inches wide. Their bite, he contended, was fatal.

"In the Kubbo-Kale valley," reported British naturalist H. S. Wood in 1935, "I saw a centipede ten inches long. Its general color was electric blue with bright coral red fangs. It was the most terrible thing I have seen in my tramps through the forest." Wood was stung by one of these Indian centipedes; he described the sensation as "exactly like that of a third degree burn."

These animals are neither snakes, insects nor worms. They constitute an independent and intermediate order of animal life. They are considered a little nearer to the spiders than to true insects. They have retained the ways of life of the ancestral worm.

Most of the centipedes are active, ferocious, flesh-eating animals. Their poison fangs are deadly to their normal prey—earthworms and insects. Some of the larger species do not hesitate to attack lizards and small mice. A bite, however painful, probably never is fatal to a human. All are land animals which creep or crawl under logs and bark. They usually remain in seclusion during the day but come out of their retreats at night when they wander over the ground and attract attention to themselves by their phosphorescence. A few have been described as sea dwellers but these do not actually live in the water. They crawl along the shore and are submerged by each tide. Some or completely blind, others have many eyes.

The centipedes are among the most repulsive of all animals, yet there are accounts of South American Indian children who drag very large ones out of the earth and eat them. Religious fanatics among North African Arabs swallow them alive as proof of their supernatural powers.

Tropical America has many varieties with varied and curious habits, like the Nicaraguan species described by Thomas Belt:

"Among the centipedes was one which had a singular method of securing prey. It is about three inches long and sluggish in its movements but from its tubular mouth it is able to discharge a viscid fluid to a distance of about three inches, which stiffens with exposure to the air to the consistency of a spider's web, but stronger. With this it can envelope and capture its prey, just as a fowler throws his net over a bird.

"Some of the other centipedes have phosphorescent spots in the head, which shine brightly at night, casting a greenish light for a little distance in front of them. I think these lights may serve to dazzle or allure the insects on which they prey."

Centipedes have been observed attacking earthworms. One may grapple with its victim for several hours before killing it. Then it sucks the blood.

A fairly familiar visitor in the southern United States is a house centipede which thrives in damp basements and sometimes invades ground floors. It is a wormlike creature, about an inch long, with fifteen pairs of long legs. In the female the last pair are twice as long as the rest of the body. The animal is yellowish grey with white bands on its legs. It is poisonous, but its jaws are weak and it seldom bites human beings. Despite the evil reputation of its race, this centipede should be a welcome guest for it feeds on cockroaches, flies, spiders, moths, and other domestic pests. It is a fast runner but often stops suddenly, remains absolutely motionless for a moment, and then darts for concealment.

THE PLANT THAT MAKES MEN DUMB

A plant now being cultivated in the newly established botanical garden of the University of Caracas may prove to be nature's greatest boon to pestered husbands and harassed mothers. It is described only under the popular Spanish name of "planta del mudo." It looks like sugar cane. According to reliable reports anybody who chews the stem is stricken dumb for 48 hours.

Other curiosities of the garden include a plant which allegedly can stimulate hair growth on bald heads and a bush whose blossoms open snow-white in the morning and turn red at noon. Here also blooms the exotic "Queen of Night," a climbing cactus with a white flower five inches in diameter which opens at sunset and closes at sunrise.

THE SCOURGE OF THE EARTH: LOCUSTS

From the days of the Hebrews prophets a visitation of locusts has been considered one of the plagues of God. A migration of millions of these grasshopper-like insects in clouds obscuring the sun leaves behind a countryside devastated as though by fire. In flight they sound like a forest fire being spread by a brisk wind. Whenever they come to earth areas of hundreds of square yards almost immediately are denuded of everything green.

In history their raids have been associated chiefly with the Near East. Quite similar creatures have caused far-reaching destruction over most of the world including the United States.

The last such phenomenon was about 1880. Since then grasshoppers have hopped, not flown. There have been some great invasions, but the insects have moved along the ground where it is easier to combat them.

The reason for the transformation was found a few years ago by entomologists. Hopping grasshoppers are changed into flying grasshoppers by heat and hunger. Grown in test cages at high temperatures and deprived of succulent green food, the insects acquired longer wings, became slimmer, and took on brighter colors.

It apparently is a curious provision of nature to preserve the grasshopper race. When on the edge of perishing, they are supplied with wings to carry them to green pastures a few hundred miles away. Lately there has been some indication that those in the western United States might again enter the flying phase in the near future. During the great drought of the early thirties there was a stimulus almost sufficient to make them undergo the complete transformation.

At present there seems little prospect that there will be another flying cloud in this part of the world. By planting cultivated crops on land formerly covered by grass, man provides good egg-laying grounds and plenty of green food.

Adequate information still is lacking on what makes grasshoppers increase and decrease. Also a mystery is the mechanism by which the harmless solitary phase is transformed into the dangerous gregarious phase. Several types occur in both phases and each can change itself into the other, altering their habits so that they attack in mass rather than as individuals.

During the late 1870s the flying clouds caused terror all over the world. In parts of Minnesota where the locusts landed they covered the ground three inches thick. Crops were destroyed throughout the prairie states.

The most remarkable incident was reported from Russia in 1878:

"A detachment of Gen. Lazeroff's expedition against the Turcomans met with a curious misadventure near the Georgian town of Elizavetopol. A few versts from the town the soldiers encountered an army of locusts about 20 miles long and broad in proportion. The officer in charge did not like to turn back, repelled by mere insects. The soldiers soon were surrounded. The locusts appear to have mistaken them for trees and swarmed by the thousands around them—crawling over their bodies, lodging themselves in their helmets, penetrating their clothes and knapsacks, filling the barrels of their rifles and boring into their ears and noses.

"The commander gave the order for the troops to push on the double-quick for Elizavetopol, but the road was so blocked that the soldiers became frightened and, after they wavered a few minutes, a stampede took place. Led by a non-commissioned officer who had espied a village a short way from the road, the troops dashed across the fields, slipping about on the crushed and greasy bodies as if on ice. They were detained prisoners by the insects for 45 hours, and on the way to Elizavetopol found every blade of grass and green leaf destroyed."

That same year a cross-continental train was held up for three hours near Reno, Nevada, by a host of locusts that covered the rails for several miles.

TREES CAN GROW SMALLER

Trees change size from hour to hour. The circumference of a tree trunk gets bigger and smaller with unpredictable perversity. For light on this phenomenon the world is indebted to Dr. John A. Small of Rutgers University.

About a decade ago tree scientists were provided with an instrument which could measure continuously the radial growth of a tree with an accuracy of a thousandth of an inch. With such an instrument it seemed plausible that it would be possible to tell just how much a tree had grown in a single day and its rates of growth in different seasons. A lot of the conclusions reached in this connection must now be discarded. The circumference of a tree certainly changes but not in a straight line. It may be bigger one day, smaller the next.

Dr. Small's experiments were carried out with the white ash. He found that circumference changes followed yearly, monthly and even daily rhythms but the changes in the same tree might vary by as much as 200 percent when measurements were made at different times. Daily variations have shown a tendency to reach maximum readings about 6:30 a.m. and sink to minimum in the late afternoon or early evening. Eccentric jumps and drops can be found almost any time.

UNDERWORLD CITIES

Seventeen-year locusts build great subterranean "cities" during their long sojourn in the earth's depths. The years underground are by no means a resting period—an episode of being buried alive. All the time the young locusts, in various metamorphoses, are busy building and eating. The eggs of the strange insects are laid during a few weeks late in summer inside twigs. From these eggs come minute nymphs, which at once make their way into the ground. There they shed their shells and grow rapidly. Their food is juice sucked from roots. They make successive mud dwellings attached to these roots. The largest observed in the eastern United States were eighteen inches below the surface. Each was a rough ball of earth about two inches long and three-fourths of an inch wide. The ball is lined on the inside by smooth mud and contains only one nymph. Every time an individual moults and grows larger it must make a new house.

When they emerge from the last of their feeding chambers, the locusts dig rapidly upward and construct a somewhat different type of dwelling some inches below the surface. These are two-chambered, with upper and lower rooms connected by tunnels five to ten inches long. These are so ingeniously constructed, according to Dr. E. A. Andrews of Johns Hopkins University, that they provide "the advantage of safety along with quick access to the surface when the proper time comes. In the shaft the nymph climbs close to the surface or falls rapidly to the bottom to escape attacks. The lining of the shaft is smooth mud a few millimeters thick. The shafts are by no means always straight or of uniform diameter, but may be sinuous and present swollen regions." In one area examined he found at the topsoil was such a mass of small stones and roots that the insects must actually have cut their way through roots. Large obstacles often were avoided by a change in direction.

"The chief implements used in making cavities in the earth", according to Dr. Andrews' report, "are the big first legs. Here, as in other legs, the end segment is used chiefly in walking and may be folded down when not needed. The second segment from the tip is used to pick off particles of earth. The third segment is the largest and, like a powerful thumb, acts with the opposing second segment as a forceps to pick up pellets of earth and small stones. The minute particles picked loose from the earth are raked together by the tip segment to make a pellet, which the forceps can carry or shove into the walls of the cavity. However, all parts of the body may come into use, for the hind legs and the abdomen may help shove earth aside and the head may carry earth plastered upon it. In vertical tunnels the animal braces its legs against the sides and, if disturbed, relaxes and drops down."

The last dwelling is large enough for the nymph to turn around inside and usually has a flattened floor. The top comes quite close to the surface without actually breaking through, leaving only a few millimeters of earth through which the insect must dig when the transmutation to an adult locust takes place. Examination of many of these tubular dwellings shows that there are no interconnections between them. Each has its own individual exit and along its course avoids contact with other chambers, although they often are very close together. This last home of the locust, before it emerges from the everlasting darkness to the world of light and quick death which is its pre-ordained destiny, is not necessarily restricted to the earth but may be contained above the surface. Aerial extensions may, in fact, be abundant and are in the form of turrets, towers, cones, chimneys, huts and adobe houses. The walls are of dense mud, not natural soil. Externally they are made of tiny mud pellets, but lined internally with the same smooth layer found in the underground dwellings.

PLANTS THAT CREATE MIRAGES

An explorer in the desolate heights of the Santa Marta mountains in northeastern Colombia, fog-wrapped and 10,000 feet above sea level, may see a flock of sheep grazing placidly among rocks ahead of him. Then, looking the other way, he may see an assembly of cowled, robed priests, apparently in the midst of some weird ecclesiastical ceremony. But when he reaches the places where he thought he saw these things there are neither sheep nor priests. He finds instead two strange varieties of the aster family, both among the real curiosities of the plant kingdom.

The vegetable sheep are bushy plants which grow on nearly barren ground near the mountain tops. The individual plant consists of thickly branched stems, about the size of a human finger, bearing many layers of leaves covered with wool-like hairs. Sometimes these leaves are so thick that the point of a pencil cannot be thrust through them. Some of the plants may be as large as a living-room sofa.

The extreme compactness of these plants and their dense covering of hairs is an adaptation to the hostile conditions under which they must live. The habitat consists of rocky slopes where the hot, dry winds of summer and the snows, low temperature and violent gales of winter expose them to a perpetual alternation of desert and Arctic conditions.

In the same general region are the monk plants, belonging to a different family, who have responded in the same way to similar conditions. Seen from a distance on a mountainside, especially through a light fog, a patch of these plants looks decidedly like a congregation of several hundred priests.

The vegetable sheep also are found in New Zealand, but there are no known intermediaries between the closely similar species growing on opposite sides of the earth.

THE OCTOPUS WORM: EVOLUTION'S MYSTERY

Worms that give birth to their own grandchildren, animals that have no digestive, muscular, nervous, glandular or excretory organs—such paradoxical creatures are the "dicyemid mosozoans", tiny worms that live inside octopuses. These little worms are among the most curious living things in nature. It is quite uncertain whether they are a step upward in evolution from the single-celled protozoans or, like some other worms, a degenerate form of many-celled animals. It might be maintained that they represent a distinct branch of the animal kingdom.

The body of a dicyemid consists of a single cell, almost half an inch long, in the form of a hollow tube, surrounded by a layer of small cells. The immediate offspring are formed and, in some cases, live their entire lives and reproduce in turn, inside one of these "skin" cells. The grandchildren break through the body of the grandparent at any place they choose, apparently without causing any wound, and live for a short time as free-swimming animals until they find an octopus whose kidneys they can enter. Then the whole life cycle starts over again.

Apparently the infestation in no way injures the octopus and the worms are of no practical importance in the world. Each kind of octopus or squid in coastal areas has its own particular species of these parasites of which about 35 kinds are known.

The worm's body contains no organs, tissues or glands in the usual sense of the word.

Before being born the larvae attain their full complement of body cells, are able to swim about, and have within them the germ cells that will give rise to the next generation. Birth is very simple. The larvae just push out, or are squeezed out, through the sides or ends of their parent at almost any point. The parent continues to develop and bear more larvae in the same manner. The number developing at any one time in the cell may range from one or two to 100 or more.

These larvae remain in the octopus as fully developed worms. But at certain times the germ cells develop into much smaller individuals, called infusorigens, hard to distinguish from large protozoa. These never leave the birth cell inside the parent, but produce germ cells of their own which develop into free-swimming creatures known as infusoriforms. These break away from the grandparent worm and from the octopus and become free-swimming animals. They are microscopic, less than a 300th of an inch long. They live from three days to a week. Here may be the borderline between single-celled and multi-celled animals—or perhaps the greatest degeneration in animal life.

THE MONSTER BEAR OF KAMCHATKA

A gigantic black bear, probably the largest of flesh-eating animals, lives in the dense, hardly explored pine forests of southern Kamchatka. This creature still is unknown to science. So far as known it never has been seen by a white man. There is, however, considerable evidence for its existence presented in a report made several years ago by Dr. Sten Bergman of the State Museum of Natural History at Stockholm, who spent two years on the Kamchatka peninsula.

Photographs have been taken of this animal's footprints in the snow. It leaves a track 15 inches long and ten inches wide. Dr. Bergman was shown a pelt of the giant bear. It was the largest bearskin he ever had seen, deep black in color, and covered with short hair in striking contrast to the long hair of other Kamchatkan bears. He also saw a gigantic bear skull, the teeth of which indicate that it belonged to a young individual.

Apparently this Kamchatkan black bear exceeds in size the Kodiak Island bear, which lives across Bering Strait and is the largest known flesh-eating mammal. The wildness of the country and its dense vegetation have protected the giant bear from naturalists and hunters. The whole land is a veritable paradise for bears who hide away in the dense thickets along the Kamchatkan rivers and subsist on the abundant salmon. They are so numerous that a native does not dare venture into the bush in summer without first shouting to let the bears know he is coming. They will keep out of a man's way if they are warned, but are likely to attack him if surprised.

The great majority of the Kamchatkan bears are relatively small animals, comparable to those of northern Europe. Some are black, but the majority are yellowish-white or light brown. The giant animal may be an extreme variation of this race, or may represent an entirely different species. He naturally is the subject of much native legendary. Some stories have been interpreted as indicating that mammoths existed within the time of man in the northern wildernesses of both hemispheres, but such a giant bear would fit the descriptions as well as would a small elephant-like creature.

If it were not for the great numbers of smaller bears, man scarcely could subsist in this country. There are, for example, no roads through the desolate land between the villages. But all along the rivers and through the forests are well-marked paths made by the bears who seem to have an engineering instinct in choosing the most logical places for crossing morasses and mountains. These paths are about the only means of human communication and eventually, if the land ever is settled, will become the roads. In the same way elephant trails in Africa and India and bison trails in the United States became the hard-surfaced highways of today. Engineers hardly can improve on the instinct of the animals.

The small bears also play an important part in the domestic economy of the few inhabitants. The thick, warm pelt is used as a bed. Out of the skin the natives make reins, snowshoes and dog traces. The meat is much appreciated. In remoter parts of the country the linings of the intestines are used for windows instead of glass. Many of the native medicines are derived from the bear.

Both among the Kamchatka natives and the Ainu of northern Japan the animal is revered as a god—the concept being that the great celestial bear out of his benevolence to men provides creatures in his own form to furnish them food and clothing.

STRANGE DENIZENS OF THE DEEP

Most fearsome of all sharks in appearance is Isistius braziliensis, found in the tropical Atlantic, Indian and Pacific oceans. It is a wine-brown colored creature with sharp teeth set in 20 rows which glow at night with an unearthly light.

"When the specimen, taken at night, was removed into a dark apartment it afforded a very extraordinary spectacle," relates naturalist F. D. Bennett. "The entire inferior surface of the body and head emitted a vivid, greenish phosphorescent gleam, imparting to the creature, by its own light, a truly ghastly and terrible appearance. The luminous effect was constant and not perceptibly increased by agitation or friction.

"When the shark expired, which was not until it had been out of the water more than three hours, the luminous appearance faded entirely from the abdomen and more gradually from other parts, lingering longest around the jaws and on the fins. The only part of the under surface of the animal which was free from the luminosity was the black collar around the throat."

One of the sea's strangest denizens is the bramble shark. It is a shark of medium size whose body is almost completely covered with short, sharp spines. This fantastic creature apparently is widely distributed through the Atlantic and Pacific, but it is not likely to come into the hands of collectors. Its general flabbiness stamps it as a deep water animal and the anomalous position of its fins indicates that it is a weak swimmer. Its spiny armament obviously is designed for protection.

Entirely harmless, it is probable, are the giant "basking sharks", which sometimes reach a length of forty feet. When encountered they rarely, if ever, try to defend themselves but attempt to escape by swimming slowly away. Stories that this monster dives when harpooned and sometimes will drag a small boat with its crew to the bottom now are discredited. Although it reigns as a monster among sharks it is not actually as dangerous as the common dogfish shark.

Perhaps the most dangerous are the so-called "carchaodons", found in most warm seas although nowhere in abundance. They are among the most powerful and voracious of fishes, but still far less frightful than their fossil ancestors. The latter were the largest of all fishes; they were probably twice the length of the largest basking or whale sharks. Some were more than 88 feet long.

COMMUNISM AMONG THE BEES

Honey bees have achieved an ideal communistic state. All the 50,000 or more members of a family—all progeny of a single queen—share and share alike. A single sample of sugar or nectar brought into the hive by a forager is participated in by all the bees. Thus all get essentially the same diet. They all acquire a common odor by which they can recognize each other. This odor constitutes a "scent language" which is the basis of the extremely complex bee social life.

These observations, based on experiments with radioactive sugar, are reported by Dr. Roland Ribbands of Cambridge University. In one of these experiments, Dr. Ribbands reports, "a marked bee is trained to collect sugar solution from a small glass tube, and when radioactive sugar is substituted the bee continues to collect the radioactive syrup quite happily. It returns to the hive and what happens to the labeled sugar can be followed quite easily. Every bee that receives some can be spotted by means of a Geiger counter. By collecting a sample of bees from the hive, one can discover what proportion of the colony has acquired some of the sugar. One stomachful can be shared among almost all the bees of a large colony. The experiments indicate that this sharing is a random affair. The sugar is passed on irrespective of the recipient's age or occupation."

Building up of a colony odor through universal sharing of the food supply enables members of the colony to recognize each other. This apparently makes little difference when food is abundant but becomes of great importance in periods of scarcity.

"At those times of the year," Dr. Ribbands points out, "when there are insufficient flowers to provide all the bees with food, they often try to steal the honey stored in other colonies. Then the ability to recognize hive mates and to distinguish them from other honey bees will enable a colony to defend itself against attempts at robbery.

"However, the honey bee community does not defend itself by attacking every invader that does not possess the community odor. Strangers are attacked only under certain circumstances. In order to investigate these circumstances two colonies of differently colored bees were placed close together, with their entrances only two inches apart, so that bees often went into the wrong colony by mistake. When good supplies of nectar were available, the intruders were allowed to enter the strange colony, but when nectar was short the strangers were attacked and thrown out, often being killed in the process.

"Production of a common and distinctive odor which enables the colony to defend itself against members of other communities is a very important consequence of the habit of food-sharing. Better sharing means better defense and so a greater likelihood that the community will be able to survive and perpetuate its kind. The habit plays the key role in the system of communication which enables the new forager to learn about suitable crops, in that the new recruit always receives a sample of the crop the colony is working. The first flight becomes a search for a crop with a similar scent. The habit enables the worker bees in a colony to be apprised of the presence of their queen. A substance derived from her body is conveyed from bee to bee in the shared food, and in the event of any deficiency in the substance they take steps to rear another queen.

"In addition, it probably helps to ensure an effective division of labor in the colony, which has to be so integrated that a suitable proportion of the worker population carries out each of the various tasks necessary for maintenance of the colony."

CANDLES ON BUSHES

In parts of Colombia candles in the form of white, wax-like berries grow on bushes. These berries produce oil of such excellent quality that it is used almost exclusively for altar lamps in Catholic churches throughout the country.

The berries grow abundantly on a jungle plant with leaves like those of rhubarb. In only one part of the country is the plant cultivated. It is a crop of the semi-hostile Paez Indians. Harvesting is somewhat difficult because the oil-containing white seed is inside a burred coat. This must be removed and the seeds placed in hot water. The oil rises to the surface where it can be skimmed off.

When it is desired to make candles a dozen or more berries are strung on a stick. Such a candle gives off a beautiful, soft light.

THE DESERT RAT MANUFACTURES WATER

All animals require water in their bodies, but some can get it without actually drinking. The desert rat which lives among the bare sand dunes of California's Death Valley, can get along indefinitely without water and with only dry barley seeds for food. In spite of this about 65 percent of its body weight is water. Most of the water is actually made in the animal's body. The rat's digestive processes extract the hydrogen contained in the barley seeds and combine it with oxygen in the air to create water.

THE CASTE SYSTEM OF THE TERMITE

The oldest civilization on earth is that of the termites. The super-organization which these blind white creatures of the dark have achieved precedes by thousands of millenia those of the ants and the bees. Termites have a far longer history on earth, being considered modifications of the ancient cockroaches who were among the first insects to leave any traces of their existence on land. Cockroaches swarmed in the club moss forests at least 250,000,000 years ago. The termite order is at least 30 million years old; some of its most primitive forms still are alive.

In most of the approximately 2,000 species of termites which have been identified all over the world there are five castes, apparently determined from birth although not so rigidly as among ants. First are the winged males and females with large brains and eyes and hard, dark shells. These depart in great swarms from the ancestral nest once or twice a year, usually in spring and fall. They are feeble flyers and depend chiefly on transportation by air currents. The majority are eaten by birds. The few surviving pairs from such a flight excavate cells in the earth or in wood and start new colonies. There is at least one king and one queen in each cell. Sometimes there are two or more pair. They remain partners for life. Both are imprisoned within the cell. Before entering it they slough off their wings, which henceforth would be worthless.

The termite queen becomes an inert, egg-laying machine, sometimes the size of a small potato. In some species she lays an average of sixty eggs a minute, or 80,000 a day. She may live as long as ten years. Thus each queen ideally produces about a half billion new individuals. Her bulk increases as much as 50-fold in adult life—about the most phenomenal growth in nature.

The second termite caste, for which there is no parallel among the ants, consists of both males and females with only rudiments of wings, less fully developed reproductive organs, and somewhat smaller eyes and brains. They presumably serve only as an auxiliary royalty, functioning in case the true rulers die. Apparently by some subtle alchemy known only to termites they can be transformed into fully functioning sexual individuals if an emergency arises.

A third caste is made up of smaller insects with extremely minute eyes and brains and barely discernible reproductive organs. Below them come the entirely unpigmented, soft-bodied workers with still smaller eyes and brains—usually, in fact, with no eyes at all. These still are potentially males and females, in distinction to any society where all workers and soldiers are female. Lowest in the scale are the big-headed, blind soldiers, also of both sexes, with barely a trace of brain.

Relative numbers in these castes differ from species to species. An analysis of an Australian termite colony accounted for 1,560,500 workers, 200,000 soldiers, and 44,000 potentially reproductive individuals.

THE SHARK THAT STANDS UPRIGHT

Monster of Gulf of Mexico waters is a shark which weights from ten to twelve tons and is from 30 to 50 feet long. Largest of its ancient family and an entirely inoffensive creature, this strange animal literally stands upright while feeding.

On a recent trip a U. S. Fish and Wildlife Service ship encountered several large schools of black-finned tuna. In the middle of each school was a large object which looked like a barrel. This object was the snout of a whale shark.

The creature kept opening its enormous mouth two or three inches below the surface. From 50 to 100 gallons of water would flow into the mouth and be strained out through the gills. This water was full of larval crustaceans, or banded shrimps, about a half-inch long.

In each observed case the body of the shark stood vertically. Why each shark should select a school of tuna and put itself almost precisely in the center of the swarming fish is a complete mystery. It does not eat tuna, except possibly very small ones. Presumably, however, it feeds on about the same sort of material as the fish. It knows there is food where the tuna congregate.

The whale shark is among the most mysterious of the larger sea animals. It is a solitary creature, seldom seen. Its tiny teeth are only about one fifteenth of an inch long and it is supposedly entirely a feeder on plankton, the minute organisms which abound in sea water.

THE DEAD MAN'S VINE

A semi-legendary plant in Colombia is the ayahuasco or dead man's vine. From it Indians make a brew which, it is claimed, is quite similar to the imaginary drug by which Dr. Jekyll split the good and evil elements of his character. When a medicine man first gulps the brew—this is an ethnological report which the botanists cannot confirm—he turns deadly pale, trembles in every limb, and the expression on his face is one of intense pain and horror. This is followed in about a minute by a reckless fury in which he seizes whatever lies at hand and starts beating the trees and ground. In about ten minutes the excitement leaves him and he falls to the earth, completely exhausted. There are not as yet any scientific accounts of the plant's influence.

THE INSECT WITH FOURTEEN LIVES

A pinhead-sized wormlike larva of a louse may possess one of life's ultimate secrets—an elixir of controlled growth.

The strange ways of life of hormophis hamamelidid—which goes through fourteen different life stages in the course of a year's lifetime—are being studied by scientists in the hope of isolating a mysterious something which may open the door of some of the greatest paradoxes of biology.

The insect is an aphis which causes galls, growths comparable to animal cancers, on witch hazel leaves. These growths result when the aphis injects into the leaf by means of a microscopic apparatus like a hypodermic needle an infinitesimally minute amount of an unidentified substance. The gall grows around and over the insect. It becomes the tiny creature's home.

The substance completely changes the nature of the plant cells. They normally would become leaf cells, highly specialized to fit into leaf growth. Now they become gall cells. Something similar happens in cancer, except that the new cell growth, having escaped from the government of the animal body, is entirely uncontrolled. The gall cells, however, still remain under some sort of control. They always form galls and they do not kill the leaf, which is necessary for their existence.

Marvelous is the life story of the aphis itself. The sequence starts with a "stem mother", a newly hatched female. She injects the substance into the leaf and the house builds itself around her. Inside this house she passes through four stages. Her structure changes completely four times. That is, she becomes in a sense four different animals, one after another. In the fourth stage she gives birth to from fifty to a hundred living young.

Each of these young, in turn, goes through four stages. In the last of these they have wings. The winged insects crawl out through a hole in the bottom of the gall. Each produces from ten to twenty young on the bottom of the leaf. Each of the young, in turn, goes through five stages. During the last they are both males and females. This is the only time the male makes its appearance in the life cycle. All the other births are by parthogenesis.

Each of the females lays eggs in the winter on the witch hazel. The buds are destined to become leaves in the early Spring. The eggs hatch a few days before the leaves appear. Each of the newly hatched aphids—all females—injects some of the house-building material into the leaf upon which she finds herself. She becomes a new "stem mother" and the strange process starts all over again.

The rapid reproduction rate might well be overwhelming to the witch hazels, and consequently suicidal for the insects, except for certain enemies which keep down the numbers of the "lice". Such tiny forms of life as larval lacewings are able to crawl through the hole in the bottom of the gall and feed on the occupants during their various stages.

University of Virginia biologists who have been giving particular attention to the aphis are interested primarily in the substance injected into the leaves. It must be one of the most potent growth factors in nature. The amount any one aphid is able to inject is indescribably minute, even though some of them make as many as 50 separate injections. The material causes the leaf cells to become larger and to multiply much more rapidly until a "house" many times the size of the aphis is complete in a few days. The structure is perfect, even including a "picket fence" of tiny hairs around its base to keep out invaders.

The substance exists in such minute amounts that thus far it has been impossible to isolate it in anything approaching a pure form. The Virginia biologists have set themselves a task requiring infinite patience over many years—tracing the increase of the amount in the salivary glands of each individual through each of its fourteen lives, and also through the eggs with which the strange life cycle starts.

The present clues indicate that the substance is a filterable virus—tiniest of living things compared with which the pinhead-sized aphis is like a whale compared to a fly.

SHYNESS CHARACTERISTIC OF GIANT RATS

Biggest of the extant true rats is the giant rat of Liberia. It is two feet or more in length and is similar in appearance to the Norway rat which infests houses all over the world. Fortunately this creature never has invaded the homes of men. It is a shy animal of the cane brakes.

NOCTURNAL POTTO

One of the weirdest of living mammals is the potto—"ghost monkey", of West African jungles. It is about the size of a squirrel, with soft, yellow fur and protruding yellow eyes which shine like malevolent witch lights in the darkness of the jungle nights. The potto is a nocturnal animal of the tree tops. Its weird, whimpering cries are believed by natives to be the voices of evil spirits. The little creature is an aberrant member of the family of lemurs, ancient offshoots of the same family from which sprang the monkeys and great apes.

WHERE TREES ARE SQUARE

A few miles north of the Panama Canal Zone is "the valley of square trees." This is the only known place in the world where trees have rectangular trunks. They are members of the cottonwood family. Saplings of these trees now are being grown at the University of Florida to find out if they retain their squareness in a different environment. It is believed, however, that the shape is probably due to some unknown but purely local condition. That the cause is deep-seated is indicated by the fact that the tree rings, each representing a year's growth, also are square.

THE LAMP THAT IS A BEETLE

The most brilliant animal luminescence known is that of the carbuncle beetles of Puerto Rico. They emit a light so brilliant that one or two inside an inverted tumbler illuminate a room of moderate size so that one can read a newspaper at night. Fields are illuminated brilliantly every night by these beetles, flying about a foot above the ground. The light is not intermittent, and seems nearly continuous. It varies from yellow to green for different species; occasionally it is yellowish-red.

RAINSTORMS OF WORMS

Rains of worms often have been reported. After a summer shower surfaces of puddles sometimes will be found covered with countless thread worms or nematodes. These worms have just come out of the bodies of water beetles and other insects, where they have developed as parasites. Before the shower the insects were dormant. These little worms in farm watering troughs led to the long-held belief that horsehairs sometimes changed into worms.

This does not, however, explain the following report in the *Levant Times*, an English newspaper published in Constantinople, of August 6, 1872:

"A letter from Bucharest reports a curious atmospheric phenomenon which happened there on the 25th ult. a quarter past nine in the evening.

During the day the heat had been stifling and the sky was cloudless. In the evening everybody went out walking and the gardens were crowded. The ladies were mostly dressed in white, low-necked robes.

"Toward nine o'clock a small cloud appeared on the horizon and a quarter of an hour afterwards rain began to fall which, to the horror of everybody was found to consist of black worms the size of ordinary flies. All the streets of Bucharest were strewn with these curious animals."

THE ICY ARCTIC WONDERLAND

Abundant and fantastic are the creatures of the shallow Arctic sea bottom. All are invertebrates—worms, sea anemones and a host of other creatures—most of whom spend their lives buried in the mud.

Some of the creatures and their curious ways of life:

Ribbon worms which, when washed ashore, literally tie themselves in knots, curl up in balls, and secrete bags of mucous around themselves.

Bright green spoon worms about three inches long. These formerly were eaten by Eskimos.

Billions of small, transparent and essentially invisible arrow worms. One species, about a half inch long, apparently is the kangaroo of the worm world.

An important element of the bottom fauna at Point Barrow, Alaska, are the lace worms. Hardly a stone in the area does not have at least one lace or moss patch.

There is a delicately peach-colored sea anemone, a bottom-dwelling animal remotely related to the coral polyps, which display an amazing phenomenon, according to a Smithsonian report by Dr. G. E. MacGintie: "When it was subjected to unfavorable conditions, such as overcrowding in a pan of water," he says, "It cast out through the mouth a translucent, white inner lining with transparent, stubby tentacles. These tentacles were tiny anemones. If conditions remained adverse more offspring were cast off, each lot smaller than its predecessor." That is, when in trouble the animal spits out babies—presumably an emergency measure for preservation of the species and a way of reproduction not hitherto recorded. Apparently the same phenomenon occurs in the sea. Partly-grown specimens of these offspring dredged from the bottom, at first were mistaken for new species. Some of these sea anemones are quite colorful—

one purplish red, one lavender, one lemon-yellow, and one with translucent, peach-colored tentacles.

Numerically the most abundant animals of the Arctic are the amphipod fleas which form an important food source for fish and seals. Great numbers live on the undersides of ice cakes from which the bearded seal sweeps them with its whiskers.

FISH THAT LIVE ON LAND

Siam and Burma are the lands of queer fish—climbing fish, stone-eating fish, hunting fish, dry-land fish, singing fish and archer fish.

In the distant geological past, life on this planet was confined to the seas. Eventually some creature belonging to the common ancestry of terrestrial animals and fish emerged from the water and over a period of countless generations, established itself on land. Something of the same general sort of development may be taking place in Siamese lakes and rivers today, with a new kind of land animal in the process of evolution. Currently, two or three species of fish are learning to live out of water for considerable periods. At least one of them appears to have reached the stage where it must breathe air to survive.

These evolving dry land fish were studied intensively by the late Dr. Hugh M. Smith, fisheries advisor to the Siamese government for twelve years. One is a species somewhat like a perch in general appearance. It belongs to a group which has an accessory respiratory organ, perhaps the beginning of a lung, situated in a cavity above the gills, by which oxygen may be taken directly from the atmosphere. The gills themselves appear inadequate to sustain life. The fish probably would drown, although the process would be very slow, if kept too long under water.

A common method of fishing in Siam is with a spade. Some fish spend as much as four months of each year buried in damp soil. Local fishermen dig two or three feet deep in the marshes for them.

THE SPECIAL LANGUAGE OF BEES

Study of bee language now has advanced to differentiation of bee dialects. Some years ago Dr. Karl von Frisch of the University of Munich established the fact that bees actually possessed a means by which they could communicate with each other and without which the remarkable organization within the swarm would have been nearly inexplicable. Their language consists primarily of signs, like that of deaf and dumb persons. Dr. von Frisch reached the point where he could get some idea of what the bees were talking about and even predict their behavior from their conversation.

Recently Dr. von Frisch has found that different varieties have quite different languages, perhaps as far apart as French and German; one variety cannot tell what another is discussing. He has gone one step further—to the discovery that the insects probably talk also in sounds that are inaudible to the human ear. The audible buzzing is not a means of communication.

"There are indications," he says in a report to the Rockefeller Foundation, "that sounds, probably in the supersonic range, play a role in their communications.

"Physiologically it would be interesting to know how they judge distance. Their dances indicate with remarkable exactness the distance between the hive and the feeding place. How do they adjust themselves to the changing positions of the sun when they use it as a compass? Apparently they have an excellent memory for time, for they seem to know that the sun at a certain time will occupy a certain place in the heavens."

Dr. von Frisch and his colleagues at the University of Munich are also making an intensive study of the insect eye and the physiology of the insect sense of smell. Previous research has shown that worker bees have a special scent gland under voluntary control. Only when a good source of nectar is found is the fragrance, evidently quite powerful and attractive to other bees, released. Then it permeates the immediate neighborhood. It is the bee language equivalent for the word "Here." When a cruising worker gets a whiff of this odor it knows there is a plentiful supply of nectar close at hand and starts a search for it.

Bees cannot distinguish red from black, Dr. von Frisch has found. This probably is the reason so few red-blossoming plants depend on these insects for distributing their pollen. Nearly all red-blossoming species depend on birds and butterflies, both of which are acutely sensitive to red. One notable exception, however, is the European poppy whose brilliant red blossoms carpet the landscape in late Spring. The German experimenter has found that these blossoms are not "red" to the bee. They possess a color which cannot be described because it cannot be experienced by the human eye. The poppy blossoms reflect a great deal of the ultraviolet light in sunshine and to this the bee eye is extremely sensitive. The color must be quite different from any of the shades at the blue end of the spectrum which are visible to man. To the bee it is probably somewhat like violet.

Even the more or less degenerate human nose can be trained to discriminate some of the bee odors that apparently have so much meaning in the life of the hive. After practising for a few months Dr. N. E. McIndoo of the U. S. Department of Agriculture was able to recognize the three castes—queens, drones and workers—merely by smelling them. With more practice he was able to make even finer discriminations, as he reports:

"The younger the workers the less pronounced is the odor emitted. To the human nose the odor from nurse bees and wax generators is much less pronounced than is that from old workers. Workers just emerged from the cells have a faint, sweetish odor, but lack the characteristic bee odor and workers removed from the cells just before they begin cutting their way out omit a still fainter sweetish odor.

"Old queens have a strong sweetish odor, while that of queens just emerged from cells is much pronounced as is the bee odor of the workers. The majority of old drones have a faint odor while every young drone has a stronger one. It is slightly different from that of young workers and is less sweetish.

"All the offspring of the same queen seem to inherit a peculiar odor from her, which becomes the family odor. Apparently each worker emits an individual odor which is different from that of any other worker.

"Of all odors, that of the hive is most important. It seems to be the most fundamental factor upon which the social life of the colony depends, and upon which the social habit perhaps was acquired."

Taste discrimination is roughly parallel to that of humans. The bee certainly can distinguish the primary tastes, sweet, salty, sour and bitter. It naturally is keenly sensitive to different degrees of sweetness, yet some sugars which are extremely sweet to man are tasteless to the insects. The same is true of such sweeteners as saccharin. The bee's sense of smell also runs parallel to that of man, both in the ability to discriminate fine difference in odors and in the thresholds of sensitivity. This appears to be a very important factor in the location of nectar-bearing flowers. However, the bee appears unable to detect an odor from any great distance. It is probably due to the sense of smell that scout bees are able to locate good feeding grounds. After marking them with their own peculiar secreted odor they return immediately to the hive to tell the others about them. The dance of a returned scout varies in intensity according to the richness of the find and the workers who witness it become correspondingly excited. If the scout executes only a feeble dance there is only a small exodus from the hive.

POISONOUS PLATTERS OF THE SEA

One of the most dreaded of all sea creatures is the venomous sting ray of which there are several hundred species distributed over the world, mostly in tropical waters. On the upper side of the tail is a saw-toothed bone dagger from two to fifteen inches long which can be driven through a man's leg. The teeth extrude a venom quite similar to that of the rattlesnake.

Largest is the giant sting ray of Australian waters. A full-grown specimen weighs about 800 pounds. The fearsome and gruesome bat sting ray of the California coast weighs up to 200 pounds and is quite abundant.

All the rays are bottom dwelling animals, leading sedentary lives on flat, sandy ground. All are carnivorous, devouring smaller fish and mollusks. Fortunately they are not very aggressive and will flee from man if given warning. Still, life guard stations along the California beaches reported nearly 400 injuries from the creatures in the summer of 1952.

OUR UN-AMERICAN FOOD

A half dozen vanished civilizations make their contributions to the American Thanksgiving dinner: onions from ancient Egypt, peas from Ethiopia, parsnips and turnips from ancient China.

Aztec, Maya, the skin-wrapped Cro-Magnon all did their part in the darkness of pre-history to make possible the plates which are loaded so lavishly. They did better than they knew. Very few new vegetables have been introduced in historic times. In many cases little improvement has been made on the products of the ancients.

The story of potatoes alone contains enough romance and adventure for a good-sized novel. Its origin is unknown but its wanderings from America to Europe and back to America again constitute a fascinating story.

Cultivated lettuce never has been found wild. It is believed to have been derived from India or Central Asia. It is one of the oldest known vegetables. Herodotus, Hippocrates and Aristotle mention it in references to Greek gardens. Chaucer notes its cultivation in England in 1340. Sixteen varieties are listed as being grown in American gardens as early as 1806.

Celery is a biennial plant native to the marshlands of southern Europe, North Africa and southwestern Asia. It long was considered poisonous and was not used as food until modern times.

The Israelites complained to Moses in the Wilderness because they couldn't have onions to which they had become accustomed during the captivity in Egypt. The cultivated onion probably originated in Afghanistan.

Pumpkins and squashes were grown in America long before white men came on the scene. Evidence of both have been found among ruins of settlements of the Basket Makers, about the earliest agricultural people on this continent. They probably came from Mexico. The Hubbard squash came to light in Marblehead, Mass., in 1855. It had been growing there for more than 50 years.

Peas are the oldest known vegetables. They are believed to have originated in Ethiopia but to have spread over Europe and Asia long before the dawn of history. They were eaten—perhaps even cultivated after a fashion—by men of Europe's Stone Age. Columbus planted some in the West Indies in 1493. They spread rapidly among the Indians and became one of the chief crops of the Iroquois.

The species from which cabbage is derived grows wild in North Africa and along the European shore of the Mediterranean. It has been cultivated for 4,000 years. Greeks and Romans grew it in their gardens. Most of the American varieties, however, originated in North Europe.

The turnip is a native of central and western China. Seed probably was brought to America by some of the earliest European settlers.

The radish is a native of China and India. It was cultivated by both the Greeks and the Egyptians. The parsnip is another Asiatic root crop. It first was planted in Virginia in 1690. Only recently has it gotten away from the home garden to become a commercial crop.

Popcorn is peculiarly American. In early Spanish writings reference is made to a ritual of the Aztecs in which "one hour before dawn there sallied forth all these maidens crowned with garlands of maize, toasted and popped, the grains of which were like orange blossoms—and on their necks thick festoons of the same which passed under the left arm."

WORMS THAT COMMIT MASS SUICIDE

An entire generation of worms commits suicide every year. Every individual casts off its own head.

These worms are a Himalayan variety of naids, fresh water animals vaguely related to earthworms. They are reddish-brown and seldom more than an inch long. The majority of the worms live with their heads buried in the mud, tail ends waving freely in the air. Upon any alarm their bodies contract leaving no signs of life.

Early in the Spring these worms literally lose these heads and die. Compared with those of most worms, their regenerative powers are quite feeble. It is believed that the decapitation is due to the fact that egg-laying is accompanied by such violent contractions of the body that the front segments are disconnected.

Every few years there is a report from somewhere in the United States or Europe of enormous numbers of dead earthworms covering the ground. A correspondent of the British scientific journal, Nature, reported in 1921: "About the middle of March I saw millions of dead worms morning after morning on pavements, roads and paths. They were great and small, young and old, of every known species and genus. They lay prone and even when they were able to reach a grass plot alive they lacked the power to burrow." The phenomenon is unexplained. Examination of the dead worms shows no unusual parasite or evidence of disease.

FISH THAT SURVIVE FREEZING

There is a realm of "supercooled life." Its denizens are deep water fish that live long and happily in temperatures below the freezing point of their blood. But whenever one of them comes in contact with even a single crystal of ice it freezes almost instantly. This strange phenomenon of marine life has been observed by biologists of the Woods Hole Oceanographic Institute.

These particular fish live at the bottom of Hebron fjord in northern Labrador. The temperature there is about 1.7 below zero centigrade. Some have been caught, brought to the surface, and then plunged into a bath of sea water cooled to exactly the same temperature. They survived for several hours. When, however, one of them came in contact with an ice crystal, it froze stiff in a few seconds. The explanation, it appears, is that these fish normally live below the depth at which it is possible for ice crystals to form in water.

Very careful experiments have shown that water can be carried far below its normal freezing point if it is kept entirely motionless and is absolutely free from minute particles of any sort which are necessary for the formation of ice crystals. This is about the condition that exists at the fjord bottom. Eventually, if the temperature is taken lower and lower, such water will solidify, but into a form far different from ice. It is noncrystalline and can best be compared with glass. But even if this happened in the Hebron fjord it would not necessarily bother the fish. Their blood presumably would turn to glass. There would be no breaking of body cells such as results from the swelling of ice crystals. After an indefinite period the animals might be brought out of the solid state, if the thawing could be accomplished quickly enough, none the worse for their experience. This has been accomplished with very minute organisms, but any techniques which might be used with higher plants or animals have not yet been discovered.

The extent of life in the supercooled world is unknown. It hardly can be confined to fish. All sorts of mollusks, echinoderms and worms also are bottom dwellers in Arctic and Antarctic waters. It's not cold, but ice, that kills.

PLANTS THAT KILL

The lethal dose Socrates was condemned to swallow by the stuffed-shirtism of ancient Athens was d-propyl-piperidine. This is the deadly alkaloid in the spotted hemlock, a common European weed which now grows extensively over most of the eastern United States. A closely related European species is the cowbane which cows instinctively will not nibble.

The devastating illness which fell upon 10,000 Greeks of the Anabasis, Xenophon would have been interested to know, was caused by andromedotoxin. This is a resinous substance common to plants of the heath family the world over. It is the poisonous constituent of rhododendron, mountain laurel and some kinds of azalgias. Honey from the blossoms of plants containing it is extremely poisonous.

When pioneers first pushed their way over the Appalachians their settlements were ravaged by epidemics of a fatal disease—milk sickness. Farms and villages were abandoned as terror-stricken settlers fled from the scourge. It was due to tremetol, a complex chemical which has been found in several plants—chiefly white snakeroot which causes the disease east of the Mississippi. When cows eat the snakeroot the poison passes into the milk.

By far the most virulent plant growing in the United States is very little known although it has caused many fatalities. This is the water hemlock or cicula—very different from the spotted hemlock whose extract was forced upon Socrates. It grows in low, swampy places nearly everywhere. When the ground is soft in the spring its roots can be pulled easily from the soil and have a pleasant odor that attracts children. It causes heavy losses of livestock.

Next in virulence of all American plants is the whorled milkweed which contains a closely allied resinous material not yet satisfactorily analyzed. It has caused the death of countless cattle.

CATERPILLARS THAT PRETEND TO BE SNAKES

There are worm-snakes, snake-worms, and wormlike animals that instinctively imitate snakes. This is especially true of certain South American caterpillars—defenseless creatures whose only security is in mimicry.

A large, green tree-living caterpillar in British Guiana ordinarily remains motionless and looks like part of a vine stem. But when the branch is shaken it rears the front part of its body and stretches horizontally. At the same time it gives a twist expanding its front segment into a bulbous enlargement with a big menacing black eyespot surrounded by a yellow ring. This it remains for a few minutes, looking very much like a poisonous tree snake that lives among green leaves.

Serpent caterpillars abound in Brazil. The best example is Leucorhampha triptolemus, a creature that hangs vertically from stems of plants. When disturbed it twists and shows a front extremely resembling the head and back of a snake. The curve of the caterpillar is just like that of a serpent. It keeps up a swaying, side-to-side movement for several seconds. The whole effect is to change what seems an innocent plant stem suddenly into an open-mouthed snake with red jaws and ferocious eyes.

ALL PLANTS ARE LUMINOUS

All green foliage gives off an invisible deep red—almost black—light. This phenomenon is one of the most fundamental processes of life. It is associated closely with the photosynthesis upon which depends all life on earth. This important discovery was made recently by biologists at the Oak Ridge laboratory of the Atomic Energy Commission while studying changes in a chemical known as adenosine triphosphate in plants engaged in photosynthesis, the formation of starches and sugars out of hydrogen from the soil and carbon from the atmosphere in the presence of light. Newly acquired knowledge about the process is paving the way to improved agricultural methods.

The biologists used extracts from the bodies of fireflies which give off a bright light when this chemical—an important source of energy in muscle—is present. Then they found that chloroplasts, the parts of plants most closely associated with the photosynthetic process, also would give off light when mixed with firefly juice and illuminated. They then made the unexpected discovery that living extracts of green plants give off a light of their own without any mixing.

The light given off by the chloroplasts now is believed to be the exact opposite of the first chemical step in photosynthesis. Light absorbed by the chloroplasts forms unstable chemical bonds within the plant. A small fraction of these chemically induced compounds recombine. The energy liberated by this process is trapped by the chlorophyll molecule, which in turn gives off the mysterious light.

It has been established that leaves, if frozen while exposed to illumination, retain their light-producing ability for several months. It also has been found that certain extracts prepared from leaves undergoing exposure to light contain substances which give off a bright light when certain chemicals are added to them.

WORMS THAT LIVE IN THE SNOW

There are jet black worms that live in red snow. They come out of their snow burrows only during the late summer evening, crawl sluggishly on the surface, and disappear at sunrise the next morning. They have been observed swimming in shallow pools that form on the surface of the great Malaspina glacier which flows down the slope of Mount St. Elias in Alaska.

Presumably during the long sub-Arctic winter these worms burrow deep in the snow and remain in a torpid state. They subsist chiefly on the microscopic red algae which give the glacial snow fields a reddish tinge. The black worms themselves are innumerable. They have been photographed covering a trail a quarter-mile long at an elevation of 5200 feet in Oregon. They are enchytraeids, relatives of earthworms. The common white variety now is raised commercially in vast numbers, on diets of oat meal and sour milk, as food for fancy varieties of aquarium fish. Both worms and insects that normally live in snow fields are black.

An investigator of the Woods Hole Marine Biological Laboratory once found a multitude of white enchytraeids in cakes of ice cut from a Massachusetts pond the previous winter. They were active when the ice thawed but all died in a few days. The same investigator kept thirty specimens of another species in a tumbler of water placed on a ledge outside his laboratory window. On a cold night the water froze solid with the worms in a tangled mass in the center of the ice cake. All but three or four were alive and appeared normal when the ice was thawed.

About 75 years ago housewives of Salina, Kansas, complained that the ice delivered from door to door was "wormy." Cakes were found honeycombed with tiny white worms, probably enchytraeids. They swam about actively when the ice thawed and infested food stored in refrigerators. All died when the temperature reached about 60 F.

Whether any worm—except possibly the most minute—can survive complete freezing is doubtful. They live in little holes that form naturally when water freezes and that are kept open by heat generated by the bodies of the creatures themselves.

THE STRANGE WAYS OF SNAILS

Among earth's deadliest creatures are cone snails which inject into their victims a poison as virulent as that of the rattlesnakes. These snail-like animals have a poison-secreting gland in the head and the venom is injected through the skin of the victim by tiny, needle-sharp, harpoon-shaped teeth. It is deadly not only to many kinds of sea animals but also to man. The poison, acting on the nervous system, may in some cases kill in several hours.

Fortunately cone-shells are timid, retiring, slow-moving creatures. They are among the loveliest of all sea shells. Most valuable is the "glory-of-the-seas" cone which is worth several hundred dollars. Of the twenty known specimens in the world, only three are in American collections. Of the 300 or more known varieties only five or six from the Indo-Pacific area are definitely known to be venomous.

The "emperor's top shell" is among the earth's most exquisite and, until recently the rarest of sea shells. This shell, about five inches in diameter, belongs to a sea snail of a genus fairly abundant during the Mesozioc geological period about 300,000,000 years ago and supposedly extinct until about eight years ago when one was found alive in a Japanese lobster trap. Thereafter the snail was seen very rarely until the present Emperor of Japan ordered that all specimens be preserved for his private collection. Fortunately his interest encouraged Japanese fishermen to keep a special look-out for the creatures and since then they have been found quite frequently. They apparently are distributed around the world in semi-tropical waters. Two species have been located in the West Indies and a new one recently has been reported in South Africa. The shells are rich golden-orange in color, highlighted with reds and salmons.

In the Smithsonian collections are specimens of the "original shell collector"—the snail that collects shells. This sea snail, widely distributed in tropical waters, has the habit of gluing to its own shell fragments of the shells of other animals, bits of coral, and almost every kind of debris it can pick up. The purpose is not known, but it may be for protective camouflage. Seen in shallow water, the creature looks like a little pile of broken shells on the sea bottom.

There is a "worm snail" that builds great limestone causeways and bridges. This is the shelled sea-snail of the Mediterranean—Termetus (wormlike). When the creature is young its shell is a regular spiral which the owner, free to move about, carries on its back and into which it can retreat when alarmed. As the snail ages the shell becomes twisted and contorted, like a tube, and is attached to an offshore rock. The animal crawls inside and soon dies. There are inestimably great numbers of these gastropods. They fix their shell tombs close together. These coil around each other to form solid masses of rock. Quatrefages, describes them in these words: "In Sicily where calcarous rocks projected into the sea I found they were surrounded by a kind of causeway which, without varying much in width, yet followed all the sinuosities of the shore almost exactly on a level with the surface of the water, filling up narrow chasms in some places and forming solid archways in others. Thus it afforded a smooth and easy path to one who did not object to having his legs washed by the waves. One might suppose the white and compact cement had been consolidated by man."

The love life of some snails is confusing to Freudians. Each animal is provided with a quiver full of arrows, located in the right side of the neck. These darts can be discharged with considerable force. They are straight or curved shafts of carbonate of lime which taper to exceedingly fine points. During the breeding season the little mollusks meet in pairs. A couple will station themselves about an inch apart and start shooting at each other. Several darts are exchanged and each finds its mark. After this love duel the two embrace and, since each is both male and female, both lay eggs. The darts presumably were first developed as defense weapons and, outmoded for service of Mars millions of generations ago, now have been turned to the service of Eros.

Showers of snails have been reported intermittently. One of the most notable took place back in 1892 at the German town of Padeborn. Late in August a great yellow cloud was seen over the town. In a few minutes it burst into a torrential rain. Afterwards the pavements were covered with water snails, all with shells broken after their long fall from the sky.

Some snails can bore holes in solid rock. One, found chiefly on the French channel coast near Boulogne, has bored holes six inches deep and an inch in diameter with a cup-shaped cavity at the bottom. The cavity is used for the animal's hibernation.

A few snails are natural barometers. They reputedly are extremely sensitive to changes in humidity. One, generally grey, turns yellow just before a rain and blue afterwards.

Snails admittedly are very tenacious of life and can endure extremes of heat, cold and dessication. Many instances have been cited, some nearly incredible. In 1846, for example, a desert snail from Egypt was fixed to a paper tablet in the British Museum in London. Four years later it was observed that he had discolored the paper in his attempt to get away. Finding escape impossible he had again retired. This led to his immersion in tepid water. The creature again came to life. He was "alive and flourishing" a week later.

There are snail harpists and even singing snails. The former were described by Rev. H. G. Barnacle, British missionary-naturalist, in a scholarly paper written in 1848: "When up in the mountains of Oahu, I heard the grandest but wildest music as from hundreds of aeolean harps wafted to me on the breeze and a native told me it came from singing shells. It was sublime. I could not believe it but a tree close at hand proved it. Upon it were thousands of the snails. The animals drew after them their shells which grated against the wood and so caused the sounds. The multitude of sounds produced the fanciful music."

The singing snails in Ceylon's blackish Lake Batticaloa were described by the British naturalist Sir Emerson Tennent: "Sounds came up from the water like gentle thrills of a musical chord or like the faint vibrations of a wine glass when the rim is rubbed by a moistened finger. It was not one sustained note but a multitude of tiny sounds, each clear and distinct in itself. On applying the ear to the woodwork of the boat the vibrations greatly increased in volume. The natives said they were made by singing snails."

VISION-PRODUCING PLANTS

Among the plants used by California Indians for food, medicine, and magic is wild tobacco. It is smoked in a hollow elder stick, about eight inches long, from which the pith has been removed. A few inhalations of the smoke early in the morning are enough to overcome the smoker so that he is unable to stand on his feet. He inhales until extreme dizziness is achieved and then he touches tobacco no more for the rest of the day. Indians can give no good reason for this concentrated form of smoking. It is simply the way of their ancestors.

A mixture of plants, the honey of bumblebees, and the red scum off an iron spring constitute a popular love charm. The mixture is placed in a buckskin bag and carried under the arm. When the favor of some particular maiden is desired it is necessary only to secure something associated with her and add it to the charm. The easiest to get is a pinch of soil upon which the lady has spat. This is used not only by lovers but also by husbands wishing to secure the return of errant wives.

Almost equally as important as tobacco in the life of these California Indians is a vision-producing plant closely related to the common garden trumpetflower and to the deadly nightshade. The leaves from the east side of the plant are smoked; this brings about a state of exaltation in which various animals are seen to come and offer their help to the dreamer. Leaves from the west side are never smoked. It would mean certain death; the Indians associate the west with death.

Much the same effect is obtained by drinking a blue-frothy decoction of the root. It not only produces visions but acts as a powerful anesthetic. It is highly poisonous, however, and only those Indians who know the proper dosage make use of it. The plant is known as "grandmother," because of its comfort-bringing qualities.

THE ABOMINABLE SNOW MAN

Mysterious beast of the high Himalayas is the "abominable snow man," so-called by natives. It is evidently a four-footed, five-toed mammal that weighs from 150 to 200 pounds and lives in family groups. This much, at least, can be deduced from its tracks in the snow, according to Dr. Edouard Wyss-Dunant, leader of the Swiss Mt. Everest expedition of 1952. He found the footprints in a snow covered frozen lake at an altitude of about 15,000 feet.

Although the tracks are bear-like, the animal apparently has a quite unbearish ability to leap from crag to crag in migrations from one high valley to another. The snow prints were first reported by Himalayan explorers to be ape-like, or even almost human, and this led to speculations that some still unknown type of big ape might have evolved in the high mountains.

The tracks, says Dr. Wyss-Dunant in his recent report to the Royal Geographic Society, are undoubtedly those of a large "plantigrade animal"—that is, one that walks on the sole of the foot with the heel touching the ground. This is the way of both bear and man. The sole of the foot is from four to five inches long by the depth of the tracks, compared to those made by men of known weights. Some smaller footprints were found, believed to be those of young animals. Three of the tracks showed imprints of claws. Small triangular markings on the heels of two of them were attributed to tufts of hair that grows on the bottom of the feet.

Tracks of one animal were followed until they came to a rock several feet high over which it was necessary for the creature to jump. On the other side imprints of three feet were found close together. Apparently the animal had landed on these three feet. The tracks of the fourth foot were some distance ahead, indicating preparations for another jump. Beyond, Dr. Wyss-Dunant picked up other trails. Three were coming out of a deep valley. The fourth came off the side of a glacier. These paths joined and thenceforward continued as a single set of tracks. The animals apparently step in each others' footsteps while they proceed in single file. This is a customary procedure for mountaineers crossing a glacier where there is danger of falling into crevasses.

Nepal mountaineers have been familiar with the mysterious tracks for years but nobody has been found who claims to have seen the animal. They call it a "yeti."

"I could find no trace of meals, nor of excrement," the Swiss explorer declared. "This confirms my opinion that the animal only passes through and does not frequent these heights. We should at least have found a place of refuge, if not a lair, if the yeti was living and hunting in the neighborhood. I rather think it passes between adjacent peaks only when, having scoured one valley, it tries to reach another. This animal is a wanderer, avoiding zones inhabited by man. It probably is not a carnivore since there is very little other animal life even in the high valleys upon which it could feed. It obviously is an animal of quite superior intelligence to subsist at such high altitudes and to have kept itself hidden from humans so long."

FISH THAT SING IN THE MOONLIGHT

There may be a fish that actually sings—that is, utters melodious sounds with a perceptible rhythm or beat which can be recorded in simple musical notation. This "singing" fish, which nobody actually has been able to identify, is one of the curiosities invariably called to the attention of visitors in the Batticoloa province of eastern Ceylon. It frequents only one deep lagoon and can be heard when the water is calm. Moonlight seems to draw the organism closer to the surface. On dark, calm nights the music still can be heard, but it seems to be coming from greater depths.

The "singing" sound at least, is a verifiable fact, according to the Rev. J. W. Lange, a Jesuit priest in Batticoloa who has tried for several years to determine what sort of an organism is responsible.

It is certain, he contends, that the sounds are made by something under the water. They are heard best when the head is held under the surface. By lowering a hydrophone attached to an amplifier into the lagoon, he was able, to record the sounds. From this record a friend familiar with musical notation was able to put them on paper.

It has been established that several species of fish in the lagoon make distinctive sounds. One, a large black fish with a yellow belly and four whiskers on each side of its face, expresses sounds like a baby's fretful crying. A large chocolate-colored fish found among the bottom rocks makes a sound "like the distant echo of a large firecracker." There is a curious little scaleless fish found in schools of 100 or more; as the school moves through the water it produces a chorus of tinkling sounds. A phosphorescent light comes from inside the throats of these animals. Among all his catches Fr. Lange has found nothing which can be identified with the singing fish, but he is convinced the music comes from a living organism.

That fish can and do make sounds now is well-known. This was demonstrated conclusively by U. S. Navy investigators during the late war. They determined the characteristic sounds made by a large variety of sea creatures whose chatter was interfering with underwater sonic devices.

BRAZIL'S VICIOUS GLOW WORM

One of the most unusual of all luminous creatures is an insect larva found by farmers ploughing damp soil in Brazil and Uruguay. It is a reddish-brown little worm with rows of green lights on both sides and a vivid red lamp on the front of its head. The red light is actually red—not white light shining through a reddish skin. Adult females of the species retain the same luminous pattern. Male adults have only feeble, yellow lights. The larva are extremely vicious little creatures, predators on white grubs which infest the soil.

GRASSHOPPERS LIKE CHAMELEONS

There is a jet-black grasshopper that turns sky-blue at sunrise. The curious creature is found on the summit of Mount Kosciusco, highest peak in Australia, where snow lingers into late summer and nights are bitter cold.

The insect is of peculiar interest because of a temperature control mechanism otherwise unknown in nature. Several animals, notably chameleons and some fish, can change color, usually to match their environment. The changes are brought about by certain hormones, released by stimulation of the eyes, which activate different color cells in the skin. But in this grasshopper every one of the outer layer of cells of the body is a color cell. On the surface are granules of black pigment, underneath granules of blue. These change places in response to temperature changes. At approximately 25 degrees C. the blue granules rise to the top, displacing the black. At 15 C. the reverse happens. This displacement can be brought about only by temperature change. Australian entomologists have in vain tried every other sort of stimulus, including illumination with various wave lengths of light.

The phenomenon probably is protective. Seemingly because it is very cold at night on the high mountaintop the black pigment absorbs and retains all the heat available. It is as if the grasshopper carried a woolen blanket. With sunrise an abrupt change takes place; and the days often become intensely hot. If the black coat were retained, the grasshopper would become overheated and probably die. The blue reflects much of the heat.

With the first streaks of sunlight grasshoppers which have slept all night at the foot of grass stalks begin creeping slowly upward. There apparently is no nervous control of the color change. Each color cell seems to act independently. The same reaction takes place in dead grasshoppers when the temperature changes, affecting even fragments of their bodies. It is possible to get a grasshopper half black and half blue by heating one end and cooling the other.

BEETLES THAT HELPED AN ARMY

During the invasion of Normandy in 1944 Army jeep drivers prohibited from using headlights of any sort, were able to follow winding country roads on the blackest nights by rows of millions of flashing green lights which outlined the roadsides.

Wingless, wormlike female beetles, (Lampyris hoctiluca, the European glow worm) were trying to attract their winged, lightless mates. Their nocturnal lovemaking as they clung to roadside weeds and bushes was a far from insignificant factor in the Normandy operations. The worms indicated not only the direction but the width of the roads, thus forestalling fatal accidents and preventing drivers from going astray into hostile territory. However, they doubtless proved of equal value to the enemy. These accommodating creatures, unknown to soldiers from across the Atlantic, should not be confused with our familiar fireflies.

WORMS IN MEDICAL HISTORY

Earthworms have an important place in folk medicine, especially in the Near East. Muzhatu-L-qylut of Hamd Allah, an ancient Persian natural history, states: "Earthworms are red worms living in the damp earth. Baked and eaten with bread they reduce the size of stones in the bladder. When dried and eaten they cure the yellowness of jaundice. In difficult labor they bring on delivery immediately. Their ashes applied to the head with oil of roses make the hair to grow."

Says a seventeenth century English medical treatise: "Earthworms are hot of nature and of them are a pressious oyntment made to close woundes; and if they be sodden in goose greece and styned it is a good oyntment for to drop into a dull hearing ear. Earthworms stamped are good for payned teeth. The oyle of earthworms be greatly commended for comforting of sinews, jointes, vaines and goute. They must be washed in white wine and the oyles of verbascum or cowslopes, of roses, of lilies, of dil, of chamomill, all sodden together. When it is cold put in your erthwormes, stoppe your glass, let it stand xl days in the sunne, then straine it. It will make an excellent oyle against ache, sciatica, goute, etc."

TOADS THAT MAKE POISON GAS

Among the weirdest of American amphibians are certain of the giant toads of southwestern United States and northern Mexico which, when frightened or in pain, diffuse a deadly gas which will kill objects some distance away.

A very large toad found almost everywhere throughout the Panama Canal Zone can squirt a poison which may permanently blind a man if it hits the eyes. Nobody would bother it except that from its skin is made of the softest and most expensive of all leather.

Most toads have skin covered with warts which are more closely grouped on the sides of the neck than elsewhere. These, together with the paratoid glands situated behind the eyes, secrete a milky, poisonous fluid whenever the animal is molested. The secretion is an acid irritant, causing pain in cuts and producing a bitter, astringent sensation in the mouth.

PLANTS THAT THRIVE ON ICE-BLOOM

There are plants that grow in ice and snow. This phenomenon—known to botanists as cryovegetation—has been the subject of intensive study at Mt. McKinley National Park in Alaska.

The plants are responsible for the strange phenomenon of ice-bloom. Ice fields at various seasons take strange colors. The plants are very minute members of the almost universal algae family which are among the most primitive forms of life on earth. They are able to extract the nourishment they require from the surface of a glacier as it melts slightly under the glare of the Arctic sun. The phenomenon has been reported by Arctic explorers for many years but until a few years ago very little was known of the responsible microorganisms. They are a striking demonstration of the fact that life has spread to all possible habitats on earth in some form or other, even to fields of solid ice.

While nobody is likely to stake out a few thousand acres of glacier for a farm, an Hungarian botanist, Dr. Ersebet Kol, has made first-hand studies of the conditions under which the minute plant organism could live and multiply, including the acidity of the ice. Concerning the Columbia glacier, one of the largest in the Alaska ice-fields, Dr. Kol reported to the Smithsonian Institution: "When I stepped on the ice, I saw for the first time a phenomenon to be seen only on coastal glaciers. The surface of the ice was covered for miles and miles with light brownish-purple algal vegetation called ice-bloom. This effect is produced by immense quantities of minute plants called Ancyclonema, a characteristic plant of the permanent ice. It can never be found elsewhere, even on permanent snow. It belongs to the green algae first found on the coast glaciers of Greenland. Since that time, the microorganism has been found in several localities in Europe, and I have found it occasionally on the glaciers of the interior but never in sufficient quantities to form the ice-bloom of the coastal glaciers.

"Here I had an opportunity of studying another striking phenomenon of the permanent snow regions of Alaska—colored snow, especially red snow. Above Valdez, around the Thompson Pass, the snowfields glitter with a reddish color in the beginning of August. The snow was red not only on the surface, but also to a depth of several inches and even in one place to a depth of two feet, caused by the presence of millions of tiny plants, Chlamydomonas nivalis. The snow on Thompson Pass looks as though it has been sprinkled with red pepper, differing in this respect from the red of other snowfields, which is usually a light raspberry red."

POISON ARROW FROGS

There is a green frog, about the size of a half dollar, that is one of the most virulently poisonous creatures on earth—but only after it has been roasted alive. It is common at the Smithsonian Institution's tropical wild life preserve in the Panama Canal Zone. When living it is quite harmless, at least to human beings although some believe it can poison other frogs. When it is roasted over a slow fire, however, a toxin is exuded from its skin which is a potent nerve and respiratory poison. It once was used by the

Choco Indians to poison the arrows with which they hunted game and Spaniards.

The poison arrow frog is a delicate creature which is confined to a narrow temperature range and probably never has reached the United States alive. A ground and tree-dwelling animal, it is quite elusive.

A close relative is a brilliant scarlet frog, a denizen of the treetop of the dense Panama rain forest. From its skin also is exuded a virulent poison. One of the two jungle canopy frogs, it is less than an inch long. Its body has deep scarlet both above and below; its feet are black and its thighs are flecked with metallic green on the rear and metallic blue on the front. It is found only on the Atlantic side of the isthmus near the mouth of a small bay where Columbus once landed for fresh water. Outside its narrow range the creature has never been seen in its gorgeous colors. In captivity it probably would die very quickly. Placed in a preservative, it quickly turns to a drab, uniform black.

The animal is a remarkable and peculiar climber. It ascends a tree trunk by a series of short jumps, catching its toes in rough spots on the bark. (Other tree frogs have suction disks on their feet by means of which they can walk up a tree in leisurely fashion.) It makes its way unerringly from the ground to its treetop home, a pool of water in the axil of a bromilead or "tank plant," a tree of the pineapple family.

THE SEAL THAT CAN "LOSE" ITS HEAD

An animal that can pull its head almost completely into its neck has recently been added to the mammal collections of the Smithsonian Institution. This is the Ross seal, one of the rarest of all the seal family in the Antarctic.

A frozen specimen captured by the Navy's polar expedition in 1956 arrived at the U. S. National Museum in Washington in excellent condition. This seal—about 8 feet long—dwells exclusively on the drifting ice pack of the Ross Sea. So far as is known it never comes on land or on the ice shelf. It apparently feeds almost exclusively on cuttlefish and squid, which are abundant in Antarctic waters. To judge by the nature of its teeth it undoubtedly is not a fish-eater. It is yellowish-green on the underside and blackish-brown on the top, the fur often being marked with pale streaks along the sides.

On the drifting pack it has fearsome enemies—notably the killer whale and the writhing, snake-like sea-leopard, most savage of the seal family—which may account for its relative scarcity. The outstanding peculiarity of the creature, probably unique among mammals, is the thick bloated neck into which the head can be withdrawn. This may be a protective characteristic although it could hardly serve the creature against its fierce enemies. On the other hand, withdrawal of the head may be a comfortable habit in a very cold climate.

THE DELECTABLE HORNED VIPER

All along the Nile and the Red Sea coast is found the horned viper which lives buried wormlike in the sand with only its eyes and the upper part of its head visible. Its horns are said to look like barley grains and to entice birds. It is found often in rodent holes. This horned viper is extremely tenacious of life. It has been kept alive in a glass jar, without food, for two years. It can hurl itself forward as much as three feet. A full-grown specimen is about 18 inches long and quite poisonous but Egyptian magicians have been seen eating the animals like stalks of celery.

FLYING SNAKES, FROGS AND TOADS

There are flying snakes as well as flying frogs and toads. Such reptiles and amphibians should be considered expert parachutists rather than actual flyers.

The tree snakes dendrolaphis and chrysopelea leap from high limbs, stretched out lengthwise and both flatten and broaden the body so that it presents a concave surface. They glide to earth slowly, at an angle to the vertical, and land apparently without injury.

Frogs of some species have enormous webs between the fingers and toes which serve as parachutes. A Brazilian tree frog has been observed to drop from an altitude of 100 feet and land 90 feet away uninjured. Since other frogs of the same size were killed when dropped vertically, parachuting must be considered a distinct trait of this particular species, developed over many generations of life in treetops.

In the course of experiments a South Carolina lizard, frequenter of bushes and fences, landed ten to twelve feet away from the place where it was dropped, at a height of 37 feet, and hopped away unhurt. It took a rigid posture when dropped, limbs outstretched and stomach taut. It fell vertically a third of the distance to the ground and then started to glide. A lizard of another species from the same habit wriggled all the way down.

EAGLES BUILD LOG CABIN NESTS

The white-headed eagle became the national bird of the United States by act of Congress on June 20, 1782. For nearly two centuries it has remained the American symbol of fearlessness and freedom. The same bird—Haleoletus leucocephalus and not the more familiar golden eagle found in the West—had been the supreme totem animal of the Six Nations of the Iroquois from whom many institutions of the new republic indirectly may have been derived.

This eagle still is fairly abundant in the fringes of forest around the Great Lakes, its fishing grounds. Its nest, almost always at the top of a tall sycamore or hickory which is dead or dying, is almost literally a log cabin. The bird sometimes uses sticks six feet long for the outer walls. It grasps large dead branches in its talons, breaks them off by sheer force, and flies away with them. A recently observed nest was nine feet high and six feet in diameter.

THE PREDATORY MANTID

Why does the "praying mantid" pray? The prayerlike pose of this near relative of the cockroach is its normal position both for seizing its victims and for defending itself.

For their size mantids are among the most predatory animals in existence. They are also among the least known of the insects. There are more than 1500 species in the world, mostly tropical. Only 19 are known in the United States which is on the northern fringe of their normal habitat. One of the most remarkable features of the mantid is its front legs, which bear sharp spines and fold in a curious hinged fashion enabling the insect to reach forward, seize a fly or some other victim, and bring it to its mouth. This is the explanation for the seeming attitude of prayer.

Mantids feed entirely on other animals, chiefly insects caught alive. Instances of small birds, lizards and mice being eaten have been reported, probably due to mistaken observations. There is no question that mature individuals of several species can handle any caterpillar, grasshopper, cockroach or other large insect that comes within its range. Their appetite is enormous. An adult mantid has been known to eat ten cockroaches in less than three hours. Bees and wasps usually have no terrors for the predators, although occasionally a mantid is stung while trying to catch a wasp and gives evidence of the injury.

Sometimes the mantid's front legs are held in a posture of sparring, rather than of prayer. More than once the sight of one of these insects "sparring" with an English sparrow or some other small animal has attracted a crowd on a city street and gotten paragraphs in the local newspapers.

The mantid usually waits motionless until its prey comes within reach but sometimes, supposedly when very hungry, it may stalk another insect. Sometimes the victim is touched lightly with the antennae before the front legs flash forward and make the capture.

These insects have developed considerable camouflage. Some tropical species look like flowers, their colors blending with those of foliage. One species varies in color from white to pale pink and has the practise of crouching among certain blossoms, the petals of which its legs and other body parts resemble. Others have arranged themselves on plants so that they look like blue flowers. Presumably bees and other flower-loving insects thus are lured to their doom. A few tropical mantids have developed a superficial resemblance to other insects of the same environment which are distasteful to birds and monkeys. Some closely resemble large ants.

There is a widespread belief that the male always is eaten by the female after mating. Sometimes this happens, but the male never is a willing victim and quite frequently escapes. The eggs are laid in groups of from a dozen to about 400. They are deposited in layers in the midst of a thick frothy liquid which soon hardens and becomes fibrous. For the most part, each species deposits egg masses of a distinctive shape.

On the whole, they probably are beneficial insects because the greater part of their prey consists of species injurious to gardens. The possibility of propagating them for the control of injurious insects, such as Japanese beetles, has been suggested because of their notoriously big appetites. It would, however, be impossible to restrict them to a specific pest. They would continue to eat about every living creature of the right size that came within reach of their claws, including many beneficial species.

FIREFLIES AS ELECTRICIANS

The flashing of a field of fireflies is an expensive show. For two generations one of the ideals of science has been to produce artificially "cold light"—radiation confined entirely to those wavelengths to which the retina of the human eye is sensitive without any energy being wasted in the form of heat or invisible light. Could the ideal be attained with the same expenditure of fuel and power as is required for light production at present the world's bills for illumination would be decreased enormously.

Actually the firefly has attained this ideal in one direction. It emits only visible light. From this point of view the firefly or any other sort of luminescent animal is very efficient indeed. A good part of the total radiation from any man-made source of light—or for that matter from the sun—is invisible infrared, observable only as heat. Possibly the firefly produces some heat in its light production but it is too little to be measured. It is safe to say that within a tiny fraction, 100% of the radiation produced is in the visible spectrum—most of it shorter wave lengths than those which produce the sensation of blue light. This is by far the highest efficiency known to science.

Chemists can duplicate the process to a certain extent. Consequently a great deal of research has been devoted to the light-emitting mechanism, physical and chemical, of the insects. Firefly luminescence is due to the oxidation—that is, the burning—of a chemical substance, luciferin. This reaction, in turn, depends upon a catalyst known as luciferase. The same phenomenon can be brought out by appropriate mixtures of luciferin, luciferase, and oxygen in a test-tube at the proper temperature.

All these experiments have shown that, considering the amount of oxygen necessary, it is a very wasteful process. It is far less efficient than most means of producing artificial light known to man—one percent compared with the 4.54 percent of the carbon filament; 17.17 percent of the acetylene flame, or 60 percent of the sodium arc light. To illuminate houses or streets with firefly light would be a very expensive procedure indeed.

Dr. N. D. Maluf of Yale University quotes a calculation that "an area of firefly light six feet in diameter on the ceiling of a room nine feet high would give ample illumination for reading or drawing on a table three feet high." This would hardly interest an illuminating engineer. The light can, however, be used in an emergency. During the Spanish-American War Major General W. C. Gorgas is reputed to have used the light from a bottle of fireflies to perform an emergency operation. The average householder would rebel at the monthly bills.

The actual light from a single firefly is very minute indeed, averaging little more than 25 thousandths of a candle power. The combined courtship efforts of a whole field full of the insects would hardly light a single room enough for sewing or reading. The insect will sometimes glow steadily with a light as low as two hundred-thousandths of candle power intensity.

Among fireflies, flashing is essentially a courtship phenomenon, yet there is no discernible difference between the quality of the light of male and female insects. What actually happens is that the flash of the female in response to the signal of the male is timed almost exactly at a trifle over two seconds. The male is instinctively aware of this time interval, so that he does not become confused with the signals of other males. In a large group of the insects the flashes of the two sexes tend to become synchronized, producing a field of light.

THE MOLLUSK VAMPIRE OF HELL

Black demon of the realm of everlasting dark is Vampyrotouthis infernalis. Most nightmarish of living animals, this "vampire of hell" has a midnight-black body about two inches long, red-brown round face on a head almost as large as the rest of the body, red eyes an inch in diameter encircled by narrow bands of pinkish-orange, rows of ivory white teeth, ten wriggling, ever-probing tentacles extending from the head. On the sides of the neck are two powerful, flashing lights each of which is a cluster of about 50 tiny phosphorescent nodules. The entire body is covered with hundreds of tiny lights.

Fortunately nobody is likely to meet this horror of an hallucination-damned maniac's ravings on a lonely road passing a graveyard at night. It is a mollusk, a close relative of the octopus and the squid but belonging to neither family, which lives in abysses of sub-tropical seas all around the world, far below the depths reached by the most penetrating green rays of the sun. Only its relatively small size and restricted habitat prevent it from being the most fearsome, loathsome creature on this planet.

The "vampire" is a living fossil, survivor out of the demonic seas of 200,000,000 years ago which found shelter from the inexorable scythe with which time mows down demons by retreating further and further into the dark. Imprints of quite similar sea animals, probably denizens of warm, shallow waters, have been found in English rocks.

Up to now about a hundred individuals have been taken from the deep sea, mostly by scientific expeditions. Of these, nearly two-thirds have come from the Atlantic off the Florida coast and near Bermuda. There are several in the Smithsonian collections. The fantastically terrible little mollusk was first taken in the Indian Ocean by Dr. Carl Cuhn of the German Valdavia expedition about 75 years ago. Until quite recently all specimens obtained have been in poor condition and there has been considerable difficulty in classifying them. The job has been complicated by the fact that the vampire apparently undergoes a series of metamorphoses which have been mistaken for different species. During the past ten years, however, they have been studied intensively by Dr. Grace Pickford of the Bingham Oceanographic Laboratory of Yale and their fearsome reality has been established beyond question.

Naturally, since the living animal cannot be observed, essentially little is known of its habits and ways of life. Certainly it is a voracious carnivore like all others of its race and preys upon every other creature of the depths in its size range. It seems to be confined exclusively to a depth of about 1,500 meters. This is the level of the sea where, for some reason oceanographers are unable to fathom, the oxygen content of the water is lowest. It goes up immediately both above and below. The vampire, apparently, cannot stand too much oxygen. Its eggs sink to about 2,000 meters where they reach their suspension level. As soon as the little mollusks hatch they rise to their natural habitat.

The vampire has powerful tentacles but its fin muscles indicate that it is a weak swimmer. It probably lurks in the abysmal darkness for its prey to come within reach of the probing tentacles. Even with its enormous eyes and its many lights it hardly can distinguish moving objects very well and presumably is not particular about what living things it eats. Its usual victims probably are fishes and smaller mollusks. It is unlikely that the creature has many natural enemies it need fear. Unlike the octopuses, its nearest relations, it has no ink sac from which to discharge a black cloud around its body for its own concealment.

CLIMBING AND FLYING FROGS

A family of frogs that climb trees, burrow and are learning to fly are the tree frogs of Mexican tropical forests. Various members of the family are at different stages in their physical adaptation to tree life. They constitute a striking example of evolution at work as a race struggles to shake itself free from one environment and conquer another despite considerable odds.

The ends of the fingers and toes of those frogs are provided with adhesive disks by means of which the animals are able to obtain a firm foothold on relatively smooth surfaces. These disks are used mainly for climbing, or for clinging to foliage and limbs when jumping. One species is both a climber and burrower. It is an extremely timid little creature and a poor climber, but it buries itself deeply in tree mosses. Another species, which seems as much as home on the ground as in the trees, deposits its eggs on the upper surfaces of leaves overhanging the water. The tadpoles, which must return to the water for their metamorphosis into frogs, simply drop off the leaves after they leave the eggs. Perhaps the most peculiar of the family is the marsupial frog, Gastrotheca, all of whose young are sheltered in a pouch on the back of the female. Some of the family lay their eggs in nests of froth attached to leaves.

One remarkable species seems to be developing the ability to fly. Its hind limbs are elongated for jumping and it has been known to leap and alight without injury from a height of 140 feet. When handled it exudes a poisonous, milky fluid which coagulates instantly, sticking to the fingers in a disagreeable way. It has a strong odor, like that of peaches, which causes the inside of the nose to itch. Experiments are described in which this animal was dropped from the top of a high water tower. It immediately spread out its limbs and, instead of dropping vertically, sailed slowly downward and landed uninjured on the ground about 90 feet away. Apparently it was able to get the best of gravity after a drop of about twelve feet. From that point on, there was no apparent acceleration in the speed of descent. A state of equilibrium was reached. Whenever one of these frogs was thrown in the air it invariably managed, after a violent struggle, to establish itself in a balanced position which it could maintain, apparently without effort, while it glided to the ground.

Within certain limits these tree frogs can change their color so that their bodies will blend more perfectly with their surroundings. One of the most widely distributed Mexican species seems to have an exceptional color range. This particular creature also is notable for its elusiveness. It exists in countless numbers, yet an explorer may hunt for weeks without encountering a single one. Such was the experience of the German naturalist, Hans Gadow. While wandering along the edge of the forest he heard what seemed to be the noise of a sawmill in the distance. As he came nearer this sound changed into a roar like that of steam escaping from many boilers, mingled with the sharp and piercing scream of saws. It came from a meadow containing a shallow rainwater pool in which were tens of thousands of large, green tree frogs. Gadow calculated that in this pool, about thirty yards square, and in the immediate neighborhood, were more than 45,000 of the creatures. The water of the pool was covered with their spawn—a minimum of 100,000,000 eggs. The next morning there was not a single frog in sight. The water had evaporated during the night and the eggs were left to be cooked by the sun.

One of the most curious of these creatures is the banana frog, whose habitat often is the upper side of a banana leaf. It is an extremely elusive creature whose color undergoes considerable change without being specifically responsive, so far has been observed, to the intensity of light. Another curious member of the family wraps its eggs in foamy lather and suspends the whole mass between leaves or blades of grass over water in such a manner that the next heavy rain washes the developing eggs or tadpoles into it. It is necessary that the tadpole stage be passed in water. Development of means to bring this about was necessary before the family could conquer a tree environment.

Another little frog spends its entire life in the leaf-formed cup of a bromelia, a plant somewhat similar in appearance to a small century plant, which grows on the branches of trees where its roots get a precarious foothold. During the rainy season this cup becomes filled with water. There the frog lays its eggs, which hatch as pollywogs.

Truly demonic are fantastic horned frogs of Brazil which devour other amphibians and small mammals. The largest of them do not hesitate to defy a human being in the mountain rain forests, their chief habitat. They are six inches long or longer and as broad as long. Some have horns on their eyelids and the tips of their noses. All have enormous mouths, so that a mouse can be swallowed quite easily. When excited they inflate their bodies like balloons and utter bull-like bellows. At other times they are heard to cry like infants.

The horns probably serve no other purpose than to add to the ferocious appearance of the animals. They are just hardened extensions of the skin, entirely too soft to be of any value in combat. All species of horned frogs are rare in collections. They seldom are seen because of their secluded habitat and their clever camouflage. They throw loose dirt over their damp bodies until they become practically invisible.

Rarest of the family are the pigmy horned frogs which have horns on both eyelids and the tip of the nose, as well as a fringe of horns around the eyes. They are beautifully marked animals.

MAD DOG CYCLES

There may be mad dog cycles. Dogs are much more vicious in June than in the so-called "dog-days" season of July and August.

The tiny poodle and the pekingese share with the big German police dog and the Italian bull rank among the 10 most vicious of domestic canines. These are some of the conclusions reached by Dr. Robert Oleson of the U. S. Public Health Service on the basis of data about dogs in the metropolitan New York area for 27 years.

During this period, Dr. Oleson's study shows there were two 5-year peaks in rabies, from 1911 to 1915, inclusive, and from 1926 to 1930. During the first period the annual average of bites diagnosed as made by rabies-infected animals was 233, compared with only an average of 78 for the previous three years for which records were available. There followed a period of 10 years during which the number of rabies cases diagnosed in biting dogs averaged only 43 a year. Starting with 1926 the curve leaped up again and in the next five years there was an average of 288 cases a year. Then came another rapid decline.

Apparently the number of rabies cases has no relation to the number of bites. These remained practically stationary at an average of about 3,500 from 1908 to 1926. There was a sudden jump to more than 7,000 cases in 1925, just before the start of the second rabies peak. But since 1930 the number of bites reported has continued to go up, in the face of rigid muzzling restrictions, until it has reached the alarming figure of 20,000. At the same time the number of rabies cases rapidly has gone down.

The same tendency toward the mad dog cycle has been noted in several European countries. It may be due to an inexplicable waxing and waning of the virulency of the rabies virus. During the peak years extraordinary efforts were made to impound all unlicensed dogs, and the decline of the waves may have been due to the lessening of the number of potential rabies carriers by this means.

Contrary to general belief, dogs are getting better tempered rapidly during dog days. The high peak of the year in bites is reached about the middle of June. Then comes a very sharp drop, which continues steadily as colder weather comes on.

No breed of dogs is entirely free from the biting tendency, but some are much more prone to it than others. The mongrel doesn't rank among the really vicious dogs and pedigree counts for nothing. The 10 breeds, in the order of frequency of their reported bites, are: German police, chow, poodle, Italian bull, fox terrier, crossed chow, airedale, pekingese and crossed German police dog.

THE AMAZING SURVIVAL OF THE OPOSSUM

The opossum, sole survivor in the New World of a primitive and very ancient family, represents an overlooked principle in evolution—survival by endurance.

How this clumsy, persecuted animal has endured through millions of generations in the midst of savage and hungry foes is the subject of a revealing study by Dr. J. D. Black of the University of Kansas.

Dr. Black examined closely the skeletons of 95 opossums in the university museum—all killed in the immediate vicinity. Thirty-nine of them gave evidence of broken bones that had completely healed. One specimen had suffered, and recovered from, breaks of both scapulae, 11 ribs, two broken in three places, and a badly injured spine. Still another gave evidence of having suffered at the same time fractures of the jaw, the scapulae, and nine ribs. Many showed evidence of ribs and scapulae broken in several places. The ability to survive such severe injuries—they would be fatal in any other animal either in themselves or because the crippled condition resulting from them would make a creature an easy prey to its enemies—illustrates the importance of the opossum's practice of playing dead.

The opossum represents an important stage in the evolution of mammals—that of the marsupials, or pouch bearers. They presumably were quite widely distributed over the earth at one time, before the emergence of the placental type of mammals to which the human race belongs, together with almost all other warm-blooded animals. They may be the ancestors of the placentals or they may represent a different line of development from the ancestral reptiles. In any event, they are considerably nearer the type of those ancient egg-laying reptiles. They are just a step beyond the egg-laying stage.

When the placentals arose the marsupials quickly disappeared from most of the earth. They were not so well adapted for survival in conflict with the more advanced, efficient type of animal. Only in Australia did they find a haven. With a single exception, they were the only mammals there when the continent first was discovered by white men. This has led to the speculation that Australia was cut off from the rest of the world before the placental races were evolved, or before they had attained such efficiency in the ways of life as to enable them to survive. There the marsupials, without competition, were able to survive and differentiate into rich fauna of the continent—of which the kangaroos are considered the most characteristic animals.

The one exception was in North and South America in the person of the lowly opossum. All the meat-eating animals which arose around the creature fed upon it if they could catch it. It was not very efficient in getting away from a pursuer. It developed no effective armor, like the shell of the armadillo or the quills of the porcupine, with which other weak animals managed to survive. It was not even very efficient at hiding. When man arrived on the scene with his bows and his guns, its last havens, the treetops, lost their small measure of security.

All the cards were stacked against the survival of the opossum, but it developed a means of its own to keep a tenacious hold on life while far more efficient creatures—beset with new enemies and changing climates—were forced to give up. The great mammoth herds, lords of the earth for a million years, disappeared. The ferocious saber-tooth tiger and the great cave bear expired by the roadside in the race of evolution. But the poor opossum had discovered the important principle that the meek shall inherit the earth—or, at least, be allowed to live in it. It became the great pain endurer and lived by submitting and gritting its teeth. It didn't fight nor hide. It merely suffered and learned how to endure suffering. This supreme ability of the opossum to recover from injuries goes a long way toward explaining its survival.

The opossum thus appears to be the prototype of a familiar class of men and women. They are frequently encountered. As children they have almost every conceivable disease. Their adolescence is a continuous succession of broken bones. Their parents despair of raising them. When they come to adult life the story is much the same. They suffer a constant stream of misfortunes, physical and otherwise. Physicians are amazed at their recoveries. And they often survive into the 80s and 90s of life while the healthy, fortunate individuals with whom they started out are left behind in the prime of life—victims of pneumonia, heart disease or accident. When the latter die the news comes as a surprise to their acquaintances who cannot understand how the strong die and the weak survive. They ponder over the paradox that strength is weakness and weakness strength. The ancient opossum might explain that paradox if it had the means to express itself.

MAMMAL PROTOTYPES OF THE "MERMAID"

The prototypes of the "mermaids" of legend are among the least known of all animals to naturalists because of their underwater habitat and their secretive habits. They are the manatees of the Caribbean region and the dugongs of the Indian Ocean. They constitute the only remaining species of the serenia, or moon creatures, distant relatives of the elephant. Both have a somewhat human facial appearance. They feed standing upright in the water, their flippers held out before them like arms. Sometimes the females hold their calves in these flippers. Seen from a distance, they have a curiously human appearance, which may account for the many reports of mermaids and mermen.

This is especially true of the dugong—a creature of the open sea, with a white, almost hairless body. It is extremely secretive and has almost never been captured alive. When one is washed ashore or caught in a fisher's net it causes superstitious fear among the natives. The manatees are not so human in appearance and are much better known.

The creatures seldom make their appearance above water in daylight. They prefer to gaze in the moonlight, and this has added to their humanlike appearance which has given rise to the mermaid legends.

One of the few persons to study the animal at close range, O. W. Barrett, an American explorer, tells us the following concerning the manatee:

"The animal still is fairly common in most fresh-water bayous, lagoons and rivers along the east coast of Nicaragua. One of the best-known herds on the Caribbean Coast inhabits the Indio River, just north of Greytown, Nicaragua. Estimates of its number vary from a few score to several hundred. The herd apparently is stationary there and does not increase or decrease to any notable degree from year to year, although the natives take a heavy toll....

"A manatee can remain under water from 20 to 30 minutes when frightened. During the daytime the slightest unusual noise, like rain falling on a tin pail or the spitting of the hunter, is sufficient to keep the whole herd submerged for hours, yet while they are grazing the hunter may go up and slap them on the back unnoticed.

"Families consisting of a bull, a cow, and one or two calves usually...merge into a herd of from 10 to 50 or more individuals living in a certain stretch of river, concentrating during the day and scattering at night. They generally graze at night, although a few individuals may be seen feeding in broad daylight. The body is held nearly vertical while grazing. The head is held well out of water, while the armlike flippers poke the grass toward the mouth. The noise made by the flapping of the huge upper lip and the crunching of the large teeth can be heard distinctly 200 yards or more away. The sound is much like that of horses grazing in a pasture. Adult manatees appear to average somewhere between 8 and 10 feet in length. Some—old females, presumably—may reach 12 feet."

A much more seclusive animal is the true "mermaid" of legend—the dugong of the open ocean. Unlike the manatee, it is a creature of the sea and seldom ventures into the fresh-water rivers and lagoons. Few naturalists ever have actually seen one of the creatures. Mr. Barrett's first acquaintance with the creature came in Mozambique, Portuguese East Africa, when some native fishermen caught in their net what they described as a "white porpoise." They were terrified and gladly presented their catch to an Italian blacksmith. This man crudely embalmed the animal, placed it in a rough coffin and freighted it to Johannesburg, where he rented a show room and made a fortune exhibiting "the only genuine mermaid—half fish, half human."

For many years mariners in the Indian Ocean and the Red Sea have told of seeing objects resembling women standing waist high on the surface. Zoologists of the Middle Ages described a "bishop fish" which had been seen standing with outstretched arms, supposedly blessing the waters. In nearly every case, it seems likely, the objects were strange water animals—the dugongs. They have a curious resemblance to human beings, especially naked women, when seen from a distance.

Nearly all mermaid stories have originated in water where dugongs are abundant. Spanish and Portuguese sailors, the first Europeans to encounter the animal, called it the "woman fish." The creature is best known to Malagasy fishermen of Madagascar who, while they prize its flesh highly, attribute to it human qualities and affinities. After capturing one the fisherman must perform various religious rites and before he is allowed to sell the flesh at a public market he must take an oath that there have been no unnatural relations between himself and his mermaid victim.

The female's breasts are roughly in about the position of those of women. She has the habit of rising about halfway out of the water and sometimes has been described as holding her baby in her flippers. Little is known of the life history and habits of the dugong. It is a creature of the shallow sea which never has survived long in captivity. It seems to share with the elephant and with man the faculty of shedding tears when it is in trouble or pain. One which was kept for several months in the Colombo zoo in Ceylon constantly was weeping. Malagasy fishermen used to torture the animals in order to collect the tears, which they sold as love charms.

Another extant member of the "mermaid" family is the manatee, found on both sides of the Atlantic in the warm, fresh water rivers of Africa and South America. Although never mistaken for a human, it is accorded considerable superstitious regard. The Kalaboi of Nigeria regard it as a sacred animal and the incarnation of a human soul. If a fisherman kills one, by accident or otherwise, he must undergo an elaborate cleansing ceremony which involves offerings before images of his ancestors and remaining indoors for three days. During this period he is rubbed from head to foot with a yellow pigment by women of his family. While the purgative rites are in progress the women sing at dawn and dusk. On the third day there is a feast on the meat, but a bit must be given to every household in the village to lay upon the shrines of ancestors.

Both manatee and dugong, and formerly the extinct sea cow of Bering Sea, are probably the closest living relatives of the elephant. They have similar brain and heart structure. The molar teeth of the mermaid family are like those of early elephants. The male dugong has tusks. There also is a great extension of the upper lip which overlaps the side of the mouth—a start in the direction of a trunk.

The next nearest relatives of the elephants are the hyraces, or conies, of Africa and Syria, best known in the form of expensive fur coats. They look and act like rabbits. A Hebrew prophet made them symbolic of timidity. Only a taxonomist would suspect these little creatures could claim any kinship to the largest of land mammals.

LIMBLESS LIZARDS AND GLASS SNAKES

A supposedly welcome guest in the underground chambers of leaf cutter ants is the amphisbaena, a nearly limbless lizard about a foot long which looks somewhat like a gigantic earth worm. These creatures, seldom seen, can be found from Brazil north to lower California and there is one isolated species in Florida.

"Those brought to me," observed the noted British naturalist and explorer of Brazil, Henry Walter Bates, "were generally not much more than a foot in length. They are of cylindrical shape having, properly speaking, no neck, and the blunt tail which is only about an inch in length is of the same shape as the head. This peculiar form, added to their habit of wriggling backwards as well as forwards, has given rise to the fable that they have two heads, one at each extremity. They are extremely sluggish in their motions, and are clothed with scales that have the form of small imbedded plates arranged in rings around the body. The eye is so small as to be scarcely perceptible.

"They live habitually in the subterranean chamber of the Sauba ant; only coming out of their abodes occasionally in the night-time. The natives call the amphisbaena the "mai das Saubas," or mother of Saubas, and believe it to be poisonous, although it is perfectly harmless. They say the ants treat it with great affection and that if the "snake" be taken away from the nest the ants also will forsake it. I believe, however, that they feed on the saubas, for I once found remains of the ants in the stomach of one of them.

"Their motions are quite peculiar. The undilatable jaws, small eyes and curious plated integument distinguish them from other snakes. These properties evidently have some relation to their residence in the subterranean abodes."

Closely related is the Florida worm lizard, rose-colored and completely legless and earless. It is about a foot long and looks so much like an earthworm that expert collectors have been fooled. A peculiarity is that it always goes down into a burrow tail first.

The Arizona worm lizard, a somewhat fabulous animal of the same family, is not, so far as is known, represented in any collection. One veteran miner told of dragging "a purple snake with two legs on its neck" from the gravel. A woman claimed to have kept as a pet for three months "a purple snake with its legs where its ears ought to be."

All these animals are in the same general family as the glass snakes of Europe and the United States. These are long, slender, legless lizards. They are burrowing animals which occasionally are turned up by ploughmen, but they often come to the surface voluntarily at night. Specimens occasionally found in daylight usually are hiding in dark recesses.

Each animal consists of apparently quite separate parts, body and tail. The body is from six inches to a foot long, according to species, and the tail may be twice as long. The animal can disengage its tail by a single twist when caught by that organ. The slightest injury or rough handling causes this tail to fly to pieces. Each piece wriggles energetically, supposedly to attract attention while the lizard itself crawls to safety in its burrow. The body does not break up and does not, as popularly reputed, come back later to gather up fragments of its tail. Instead it grows a new tail, always smaller than the original, from the stump.

THE ONLY BUG IN THE SEA

Only one group of insects has taken to the sea—the small, gray long-legged water striders. Unlike fresh water relatives of the same genus, these have permanently lost their wings. They have no further use for this means of movement in the ocean.

Great numbers have been found floating and swimming in the open sea around Pacific islands. Both nymphs and adults sometimes are blown onto the beaches by strong winds. They are awkward on land, seek shelter in any depression in the sand, and fall easy prey to birds and the multitude of ghost crabs which glide over the sands after dark.

On the surface of shallow water the insects are found in groups of hundreds of thousands. Apparently they feed on plankton which rises to the surface at night. They themselves are not eaten by fish. This is probably due to scent glands which secrete a strong odor which is repellant to the ever hungry vertebrates.

In small embayments are found enormous numbers of one type of water strider, the female of which is less than a twelfth of an inch long. The male is considerably smaller and rides on the back of his mate to ensure that the two will not be separated by wind or tide.

Insects are by far the most abundant of all land animals; the reasons why only one genus has invaded the sea have been the subject of much speculation. On the continents, insects are found in salt water lakes where the saline concentration is much greater than in sea water. Other types live in torrential streams and waterfalls where they get much rougher treatment than would come from wave action. There are two probable reasons for the failure to invade the ocean. One is the fact that no insect ever has been able to live in very deep water. The "bug" race has evolved a special breathing mechanism admirably suited to life on land but rather poorly adapted to life under water. Besides, the seas have been taken over almost completely by the remote relatives of the insects, the crustaceans. These include, besides crabs and shrimps, the superabundant copepods, the "lice of the ocean." Invaders from the land never have been able to compete with them.

A CROCODILE WITH LIFE AFTER DEATH

There is an animal that can bite—it might even slash off a man's arm—after it is dead. Alive it is relatively inoffensive. Being killed makes it positively mad.

Its uncanny ability to bite half an hour or more after its neck has been broken is a major risk for followers of one of the most adventurous of professions—the jungle crocodile hunters. Their story is a saga paralleling that of the Antarctic whalers who first told of Moby Dick. One of the most expert of them is Dr. Fred Medem, Smithsonian collaborator and professor of zoology at the University of Bogota. He has twice been bitten painfully by "dead" reptiles.

The animal is the caiman, smaller than either alligator or crocodile and probably more closely related to the former. Its hide, like that of its two fellow crocodillians, is valuable for leather and during the past few years it has been pursued close to extinction by professional hunters in Colombian and Brazilian jungles and lagoons. Dr. Medem is an eminent zoologist. He doesn't believe, of course, that any animal that is completely dead can bite off a man's arm, but he is hard put to explain what he himself has experienced. He thinks that part of the caiman's nervous system which activates its snout and mouth is somehow disconnected from the rest and does not die at the same time. Thus the dead reptile has no consciousness when it bites. It is a reflex action of one small segment of the nervous system that somehow is not completely dead.

There is only one way to be safe for an indefinite period after the caiman is killed. That is to chop a hole in its neck and run a pointed stick into the medulla oblongata, the reflex action center at the base of the brain. When this is destroyed the ability to bite is lost. One can proceed to skin the animal without fear of losing an arm or a finger. Ordinarily this reptile will not attack a human. It lives on smaller animals—wild and domestic pigs and the pig-like capybaras—that venture into the jungle rivers.

Dr. Medem has recently discovered a curious new sub-species of caimans confined, so far as known, to the upper reaches of the Apaporis river, a tributary of the Amazon. It is much more crocodile-like in appearance than the rest of the family, with a very long, narrow snout. The others have broad, flat snouts. It retains prominent bony ridges over its eyes—one of the most striking characteristics that distinguish the caimans from both crocodiles and alligators.

A much more dangerous animal is the Orinoco crocodile, a large reptile which lives only in the Orinoco and its tributaries and has a taste for human flesh. The creature is especially dangerous to bathers and to women doing their washing in the rivers. This is one of the two species of these dreaded reptiles known in South America. The other is a smaller, less aggressive creature of seashore rivers and lagoons. The inland species now is quite close to extermination. Until recently it was pursued by both German and French companies of professional crocodile hunters. Now they have given up because the profits have become too small for the risk.

The technique for hunting caimans and crocodiles is strikingly like that of the whale hunters and just as dangerous. The hunter goes out on the river with a boat at night. The boat carries searchlights which move over the surface of the water. Here and there appear glittering red and yellow spots. The red spots are the eyes of crocodiles, the yellow ones eyes of caimans. The boat is propelled by jungle Indians who have developed the ability to paddle noiselessly. They row to within about two yards of a pair of glittering eyes. Then the hunter throws his harpoon, equipped with a special aiming apparatus. He has developed skill in hitting precisely the right spot, judged by the position of the eyes. For a crocodile he aims at where the neck should be, for a caiman at the flank. The neck of the latter reptile is protected by heavy scales. A gun never is used. The wounded reptile simply would dive into deep water where its body could not be recovered. After the harpoon, with a rope attached, finds its mark there is a terrific struggle as the reptile tries to get into deep water. The caiman finally is "killed" by chopping through its spinal cord with a machete. That is, everything is dead except the brain and the snout. The spine of a crocodile is broken by a blow with a large ax just behind the shoulders. It stays dead.

The caimans migrate overland from lagoon to lagoon during the dry season. When at last they find water they dig holes in the mud and sleep until the heavy rains return, when they emerge and resume their normal ways of life. Quite exciting stories are told of persons who happen to meet migrating bands of these "barbillos", creatures about three feet long. Ordinarily they will not attack humans but they will not hesitate to do so if they feel they are threatened. Once one of them gets a grip it is almost impossible to break away unless one happens to have a machete.

THE SALAMANDER THAT LIVES LIKE A WORM

There is an animal related to the salamander and the frog which looks like a gigantic earthworm and lives an earthworm's life. It is seen so rarely that probably not one person in a million is aware of its existence.

It is the caecilian, a very ancient creature forming the third branch of the order of amphibians which were probably the first back-boned animals to establish themselves on land nearly 300,000,000 years ago. There are about fifty species. Caecilians are found in most of tropical America, Africa and Asia. They range in length from a few inches to nearly a yard. The larger ones might be mistaken either for titanic earthworms or small snakes. In the physical structure are combined features of both salamanders and frogs.

These amphibians spend all their lives burrowing in the soil. They live chiefly on earthworms and come to the surface only for brief intervals after heavy rains. They usually are seen only by farmers who uncover them while ploughing, or digging ditches. Since they are so easily mistaken for snakes they are avoided, although they are entirely harmless. They have sharp teeth but make no effort to bite when handled.

Most of the caecilians are egg-layers, the large eggs being attached to one another like beads on a string and then wound up in a ball. This is incubated by the mother who coils herself around it. The burrows where the eggs are laid are always on a stream bank since the young, like those of all amphibians, must pass part of their development stage in water. These amphibians probably are fairly abundant animals. Owing to the subterranean life they are nearly, perhaps in some cases completely, blind.

The amphiuma, a species of salamander, also is often mistaken for a snake. It spends most of its life in rivers buried in mud, where it lives on larvae and on fish eggs. Since it is an air-breathing creature it must come to the surface frequently to breath.

The amphiuma has rudimentary legs, almost microscopic in size. This fact alone is enough to differentiate it from the snakes, who always are legless.

This curious salamander is seldom encountered and is barely mentioned in standard textbooks of natural history. Confined to the southeastern United States, it often is considered a highly poisonous animal. Actually it is harmless. Very rarely one is caught on a fishhook. It is so slippery that it is almost impossible to hold in the hand.

The creature has some relatives which are not so secretive in their habits and are much better known. One is the giant salamander of China and Japan, the largest and most active of the race. It makes its home in crevices under rocks in running streams. Another is the "mud puppy" or "hell bender" which sometimes gets on the hooks of fishermen in muddy streams.

The amphiuma is a degenerate member of the family. It has almost lost its legs. It still retains its eyes, but these have become very small. The animal can have very little use for them.

In India is found a wormlike caecilian, Ichthyopis, which lives under stones and burrows after the fashion of earthworms. Superficially it differs from an earthworm by its darker color. Its body is coated with slime and it leaves a trail of mucous behind it when it crawls.

The earth snake Silybura is found in the same region. It usually is mistaken for a worm, especially by birds to their own discomfort and sometimes disaster. It ties itself in loops around a bird's feet and these loops are quite difficult to loosen. Among natives there is a superstition that if it coils around a child's finger the only way to get rid of it is to amputate the member.

THREE-EYED LIZARDS OF NEW ZEALAND

Among sun-baked rocks on barren islands off the New Zealand coast basks a solitary survivor of the days before the dinosaurs. It is earth's oldest back-boned inhabitant, a fugitive in time from nature's harsh law of the survival of the fittest—the tuatera, or three-eyed lizard. Its big, dreamy hazel eyes have watched the procession of the ages for 300,000,000 years— the beginning and extinction of the dinosaurs to whom it stood in about the relationship of a great uncle, the coming of birds and mammals, milleniums of famine and milleniums of plenty, the shattering and crashing together of continents. It has survived while all its contemporaries of the earth's ancient days have died, largely because it has been willing placidly to watch the parade pass without bothering to take any part in the tumult and shouting.

The feature of great interest about the tuatera, both popularly and scientifically, is its third eye. This third, or pineal, eye is closer to its original form in the tuatera than in any other living creature. Just after the little reptile is hatched the organ appears as a dark spot under a film of thin, semi-transparent skin. In a baby tuatera it becomes a small knob on top of the head. Thick, opaque skin covers the eye in the adult reptile and it is difficult to distinguish. Anatomists doubt whether the animal actually sees with the pineal eye any more. The fact remains that this organ can be distinguished easily and that it retains, in degenerated form, the characteristics of a seeing eye which has nerve connections with the visual cortex at the back of the brain. Moreover, when the third eye of an infant tuatera is dissected there is clear evidence that it once was a double organ.

The tuatera is about two feet long from its snout to the tip of a crocodile-like tail. It has a scaly skin with a row of spines along its back. Its large hazel eyes are its most conspicuous feature. They have a soft, dreamy expression, and they never appear to blink. There are no external ears, but the sense of hearing is highly developed. One way of drawing the creature from its burrow is to play a tune on almost any instrument.

It does not dig its own holes under the rocks. Usually it shares the burrow of a black-and-white petrel—known in New Zealand as the mutton-bird—and it remains there even when the bird incubates its eggs and feeds its nestlings. Apparently a mutually satisfactory arrangement has been reached between petrel and lizard. The former usually are in their nests only at night. The tuatera spends most of the night away from home, hunting for the insects which are its favorite food. Occasionally, it has been observed, a host will become tired of his persistent house guest and try to evict it. In such a case the tuatera never puts up a fight. It leaves placidly and tries to find some other petrel with whom it can share quarters. If this search fails it will, as a last extremity, scoop out its own burrow, although apparently such labor is against its deeply fixed principles of making no effort which possibly can be avoided.

The lizard goes to sleep about the middle of April, the beginning of winter in New Zealand, and wakes late in August, when spring is well underway. Then for seven months it grows fat on insects.

The creature is reportedly capable of living for 500 years and more. It shares its longevity with its distant relatives, the great turtles. Its long life, during most of which it continues to breed, doubtless has been a major factor in its racial survival.

The ancient reptiles were plentiful when white men first came to New Zealand early in the last century. The Maoris regarded them with superstitious awe and avoided them as much as possible. But early British settlers and their dogs used to kill the inoffensive creatures for sport. This was the first active enmity the tuateras ever had known. They saved themselves by withdrawing to the barren islands and becoming even more seclusive in their ways of life. Thus they clung to a thin thread of existence until an enlightened government threw the protection of the law around them.

Today the three-eyed lizard is probably the world's most rigidly protected animal. The New Zealand government has placed all sorts of legal restrictions on hunting or capturing it, and to kill one would be a major crime. For that matter, very few persons living ever have seen a tuatera. It stays in seclusion most of the time. There is a single specimen in the zoological park at Wellington. When a party from a Byrd Antarctic expedition visited there they were told that the lizard had not been seen for several months and that it was highly improbable that it could be lured out of hiding. One day it would appear of its own volition, take a philosophical look at the twentieth century, eat a few flies, and retire to its lair under some rocks again. Here probably is the secret of the race's longevity. The little lizard has spent most of its time sleeping. It has existed with the minimum of effort. It has been satisfied with its lot and, above all, it never has gotten in the way. It has been observed, for example, that one of the creatures never climbs over even the smallest obstacle. It always will walk around.

PRODIGIOUS FERTILITY OF INSECTS

The capacity of insects to reproduce is almost incalculable. A single over-wintering house fly theoretically might have 5,598,729,000,000 descendants in a single year. It has been calculated that a single cabbage aphis, which weighs less than a thirtieth of an ounce, might give rise in a year to a mass of descendants weighing 822,000,000 tons, about five times as much as all the people in the world. Fortunately nearly all insects have an enormous mortality rate.

THE LIZARD THAT RUNS OUT OF ITS OWN SKIN

There is an animal that can get out of its own skin. It is a little brown lizard, a gecko, which lives in native houses on the Palau Islands in the South Pacific. This creature, about six inches long, is closely related to the house geckos, which are found throughout the tropical Pacific islands and as far north as Florida in the New World. The Palau species is almost impossible to capture by hand.

Grabbed by the tail, it immediately sheds that organ. This is a rather common practice among certain lizards and apparently brings little inconvenience. A new tail can be grown. But as soon as a hand is laid on this particular species it immediately and literally "runs out of its skin." This is done with lightning-like rapidity. The would-be captor is left holding the animal's empty skin. All the rest of the lizard is running away, presumably seeking a hiding place.

This "running out of the skin" is a far different phenomenon than that of shedding the skin by various reptiles, which always takes place after a new skin has been formed underneath. The gecko just abandons its skin altogether. It flays itself alive. Escape in this way apparently is suicidal in most cases. That it ever could grow back a complete skin is highly improbable.

HIGH LIVING IN THE HIMALAYAS

The highest land-dwelling animals on earth are small, black attid spiders. They live in islands of broken rock on Mount Everest at an altitude of 22,000 feet. This is far above the line of perpetual snow and nearly a mile above the last vegetation. Since there is no other living thing near them, they have to eat one another for sustenance. Presumably their ranks always are being repleted by new arrivals from below.

Highest of all living things are red-legged, black-feathered choughs, birds of the crow family. A lone chough has been seen in the Himalayas at 27,000 feet. There is an intimate association between these birds and mountain sheep. The chough sits on the sheep's back and searches its hair for insects. The sheep seems to like this attention and stands still while the exploration is in progress.

Another bird-animal association at high mountain altitudes is that between mouse hares, rabbit-like animals about the size of large rats, and finches. The hares live in burrows and usually are seen feeding at the entrances or running from hole to hole. Both hares and birds are seed eaters.

Wild sheep and mountain goats in the Himalayas struggle up to about 17,000 feet. There are small, wingless grasshoppers at 18,000 feet. A few bees, moths and butterflies are found at 21,000 feet.

BARKING SPIDER MONKEYS

Barking spider monkeys that fight off unwelcome human invaders are dominant animals in the "green mansions" of Panama jungles. They live in semi-nomadic troops, each of which occupies a fairly restricted area of the forest, sometimes overlapping slightly with areas of other groups. Within their territory members of a troop wander freely, but their activities tend to center around food and lodge trees.

In reporting on his observations of their activities Dr. C. R. Carpenter of Columbia stated: "Almost every night the group slept within earshot of camp. For eight successive nights they returned to the same group of trees. Throughout the day the troop travelled, in general, over the same routes from one food tree to another and from favorite places in the deep forest where the midday siesta occurred. Several other groups were regularly located in their own particular home areas."

The monkeys resent intrusion of their territories by anything that looks like another monkey, such as a man. When approached they start barking. The usual terrier-like bark of great excitement may change to a metallic chatter repeated with great frequency. When males, and sometimes adult females are approached closely they growl in a strikingly vicious manner. Typically they come to the terminal ends of branches, often within 40 to 50 feet of the observer, and vigorously shake these branches. Both hands and feet may be used while the animal hangs by its tail.

Throwing of branches is a conspicuous part of the reactions to men. Quite frequently they break off and drop limbs close to the intruder. Green branches sometimes, but most often large dead limbs weighing up to ten pounds may be dropped. "This behavior," according to Dr. Carpenter, "cannot be described as throwing although the animal may cause the object to fall away from the perpendicular by a sharp twist of its body or a swinging circular movement of its powerful tail. This dropping of objects from trees may be considered as a defensive adaptation arising from the more generalized habit of shaking branches. A significant variation occurs when the animal breaks off a limb and holds it for a time—from a second to half a minute—before letting it fall."

Normally the monkeys travel along the upper surfaces of limbs, using all four feet and carrying the tail arched over the back. When crossing from one tree to another they use their powerful tails to support themselves from limbs. During such movements hands, arms and tails are used at the same time to make contacts with supports. The monkeys have a strong tendency to keep their heads upward. Therefore, when coming down a perpendicular limb, vine or tree trunk they go backwards rather than head foremost. They frequently make long jumps outward and downward, covering at times more than thirty feet

THE INSECT THAT IS BORN PREGNANT

Among nature's weirdest tricks is the strange phenomenon known as merokinosis, reported for a single family of almost microscopic insects. The little creatures are fathers and mothers before they are born. They are a species of mite which infests grass. They belong to a family which, almost alone among insects, gives birth to living young.

Nearly all insects are egg layers. The eggs, usually deposited in enormous numbers, hatch outside the body of the mother. Then the individuals go through a series of metamorphoses—nymph, larva and the like—before reaching their own reproductive maturity.

These grass mites, however, are born fully adult animals. A sack on the body of the female swells until it is about 500 times the original body size. It is filled with eggs and a nutritive fluid. Within this sack the eggs hatch and the new generation passes through all the ordinary stages of insect metamorphosis. Finally, when they are fully mature, the mother dies, the sack breaks, and the host of new mites emerges.

It was long thought that the mites were striking examples of parthenogenesis, or asexual reproduction. Females isolated as soon as they were born gave birth to large numbers of young. Parthenogenisis is not uncommon among the lower animals. Invariably however, except in this one case, all the offspring are of one sex. The supposedly virgin birth families of the mites contain both males and females in various proportions.

BULL-DOG ANIMALS

A repressed tendency towards the bulldog face apparently is deep-seated among mammals. Foxes, cattle and pigs with bulldog appearance have been reported. In three species of dogs—the bulldog, pug and the pug-nosed dog of ancient Peru—this characteristic is dominant. It could have been caused by a pronounced shortening of the rostral portion of the skull due to the failure of facial bones to develop.

FORESIGHT OF KANGAROO RATS

A recent report by Dr. William T. Shaw tells of observations of giant California kangaroo rats whose food consists largely of the seeds of pepper grass. The seeds are gathered busily all day and stored in shallow surface caches where they are dried by the dust and heat of the sun. During the night, the animals work busily removing the dried seed to much larger chambers deep underground where it is to be stored for the winter. In some way the highly intelligent animal has learned the secret of preventing mildew. Only a few other animals have mastered the same technique; the beaver and cony dry their twigs in the sun before storing them.

THE PRIMITIVE PROTURANS

The proturans—blind, wingless minute bugs found under bark and in leaf litter—are earth's most primitive insects. They are seldom seen and when they are noticed are likely to be mistaken for larvae of some other insect. So obscure are the creatures that they were not discovered until early in the present century. They are about a twentieth of an inch long, yellowish, and covered with a protective shell of chitin. Sluggish and slow-moving proturans have three pairs of legs, only two of which are used for locomotion. The front pair is held up in front of the insect as it moves. These legs apparently serve the purpose of the antennae found in all higher insect orders. They are provided with primitive sense organs of touch. These little creatures presumably represent one of the earliest stages in insect evolution.

AIR-CONDITIONED HOMES OF BEAVERS

Air ventilation of homes appears to be an engineering accomplishment of beavers. "The beaver hut seen from the outside," according to Sigvald Salveson of Aamli, Nowayd, "appears to be so tight that it seems astonishing that the occupants can get sufficient air. In winter, when the lodge is covered with snow and ice one would not think it possible that the animals could live in apparently air-tight dwellings. Near my home is a small lake where a beaver built a dam and a great lodge. In the outlet of the lake the water was still open and I noticed the footprints of beaver on the thin ice just beyond. Twigs and small trunks were dragged to the open water, where the animals sat on the edge of the ice and took their meals. A fox had his usual track over the lodge.

"More and more snow fell and the hut was more and more hidden under the white blanket. Sometimes I noticed that the fox had gone to the top of the dome and evidently sat there for a while. Near where he had sat was a hole in the snow about half a foot in diameter and with thin ice around the edge. I found that the hole widened downward and ended on the roof of the lodge. At the bottom the hole was at least two feet in diameter and its walls were hard as ice. From this hole or chimney rose warm steam, and the twigs and mud on the roof felt warm and damp to my hand."

THE DEMON OF PUERTO RICO

In deep sunless ravines of Puerto Rico's Pandura mountains dwells the demon frog. It is a ghostly voice from mountainsides strewn with great, decomposing granite boulders and so thickly covered with tropical vines and bushes that it is almost impenetrable to man. Until twenty years ago it was only a voice, for none of the strange little creatures ever had been seen. The mere sight of the animal, according to many of the natives, would be fatal.

"One might as well try to bribe a mountaineer to catch a ghost as a guajone. There is a strange quality in the voice which probably is largely responsible for the superstitious dread of the mountain people," according to Smithsonian Institution biologist Gerrit S. Miller, Jr.

"It is strange enough when heard from the surface," Miller reports, "but it becomes even more strange after one has climbed down into the irregular and dangerous openings, which prove to be much larger and more cavernous than the surface appearance, with its dense and deceptive covering of vegetation, could lead one to expect. With flashlights the frogs are easily found and caught as they crawl slowly over the damp, but not slippery surface of the granite.

"To the natives they are objects of dread. One man said they were about a foot long and armed with frightful teeth. Another assured me that anybody who saw one would die shortly afterwards. No offer of money could induce the boys or men to go into the cavities in search of them."

The little creature is fantastic in appearance, chiefly due to its large protruding eyes. The edge of the eyelid is white, making a thin white line around the eye. The iris is back and gold. The skin is light brown above and nearly white underneath, but some specimens have blotches of yellow which add to the weird appearance.

Living as they do in the semi-darkness of mountain gullies, little is known of the life history and habits of these strange creatures. The most notable characteristic of several specimens kept alive for observation was the peculiar singing in a liquid note repeated six or seven times. It can best be imitated by whistling. This singing is believed to be part of the courtship behavior of males.

The demon frog has been given the scientific name of Eleutherodactylus cooki. It appears to have been especially adapted for life among the boulders of its restricted habitat.

MAN-MADE PLANTS

At least a half dozen species of plants are man-made. They are hybrids which can transmit their basic and unique characters to future generations.

The fact that what long was considered an impossibility in the plant kingdom has been achieved is revealed by Dr. H. Bentley Glass, professor of biology at Johns Hopkins University. With newly developed techniques which make possible the doubling of chromosomes, bunches of genes which are the units of heredity, the creation of species may be just at its threshold and man may take over control of evolution.

The definition of species, after all, is the ability to produce offspring with the major characteristics of the parents. The first successful attempt, Dr. Glass says, was by a Russian geneticist about 30 years ago. He crossed a radish and a cabbage and produced a "rabage." When two rabages were mated they produced seed which sprouted into other rabages.

Unfortunately for the man who had been the first to cross one of the great barriers in biology, the rabage was a pretty poor specimen. It had the prickly, uneatable leaves of the radish and the poor root system of the cabbage. Russian agricultural authorities had been led to expect great things. They were bitterly disappointed that the new vegetable did not fit into one of the five-year plans. The geneticist was not heard of again and it is generally believed that he was "eliminated" as a reward for one of science's greatest achievements.

Creators of new species have fared somewhat better in other countries, especially the United States, but they have not fared too well anywhere. In practically every case the new species they have created have taken over the worst characters of the parent species. They have been of no commercial value. It is likely that about the same thing has happened in nature throughout the milleniums.

But bad may be good. It all depends on the environment into which the new species is born. Under the right circumstances, the rabage might have superseded both radish and cabbage. That is, it might have been adapted to a change in environment in which both parent species would have become extinct.

Although no new animal species has yet been man-made there seems no overwhelming reason why this should not happen with some of the new chromosome-doubling drugs. However, a new kind of man is not likely. Among higher animals the mechanism of heredity is very complex indeed. It isn't likely to happen in nature, in the face of atomic radiation. It has been calculated that normally there is one human mutation per generation for each 50,000 individuals. The high probability is that this mutation involves a recessive, or hidden, gene. Its effects do not appear in the population until two persons carrying the same recessive are mated. About 999 out of 1,000 recessive genes are "bad" and in due course will cause the extinction of the line in which they appear. In the long history of the race it is likely that everybody has fallen heir to one lethal gene, but it may be a long time making its appearance in family lines.

Most of the genes in any given population, good or bad, are so hidden that it is practically impossible to predict what the offspring of any particular couple will be.

The recessive genes have vastly increased through the operation of human "melting pots" all over the world in the last few generations. One result is that minority races tend to become absorbed in majorities. Thus the relatively small American Negro population, without any further intermarriage but purely through the cropping out of recessives already received from the white majority, will be entirely amalgamated in the more numerous race in approximately 2,000 years.

Genetics is getting into the hands of scientists tools which can speed up the natural process of change about 1,000-fold and this may result in either good or evil. The good side is well illustrated by hybrid corn—a plant which cannot be considered a new species. This lately has been carried to the point where corn with much more sugar in its stalks and only six instead of twelve feet high can be produced.

THE GREAT SEAL MIGRATION

The great annual northward migration of the seals is one of the most remarkable phenomena of animal life. It seems to be without organization and without leadership. Yet toward the end of March each year the hundreds of thousands of cow seals and pups scattered over thousands of square miles of water start at about the same time in three great groups bound for three specific places. It has been the same for centuries, perhaps milleniums. Each animal moves at about the same rate so that all arrive within a few days of each other. Unlike birds, they do not move in compact masses. Three great herds exist.

The American herd of about 1,500,000 is by far the largest of the three. It goes straight to the Pribiloffs, where it goes ashore on two almost barren islands—St Paul and St George. The Japanese herd, numbering about 40,000, makes for Robben Island, off northern Japan. The Russian herd, now estimated at about 200,000, goes to a few rocky islands of the Commander archipelago off Kamchatka.

The moving herds consist almost entirely of females and young. The bulls winter further north, tend to be solitary during the winter, and precede the cows to the summer homes. The breeding season lasts for about two months. During this time the bull never eats or touches a drop of water. He never leaves the land. He arrives sleek and fat from the ocean pasture and is able to survive entirely on stored energy. This keeps him alive, even when he fights scores of terrible battles with younger rivals. Towards the end of summer he naturally is a sorry looking creature.

One day, actuated by some common impulse, cows and calves depart. Then the bulls, their arduous labors of race propagation over for ten months, draw back among the rocks for a long rest.

THE MAGIC BARK OF THE CINCHONA TREE

The shadow of a pale Spanish lady, dead for almost three centuries, has returned to the dense rain forests of the western slopes of the Andes.

The shadow is that of the Countess of Chinchon, wife of the redoubtable Don Luiz Geronimo de Cabrera Bobadilla y Mendoza, colonial viceroy of Peru. She was dying of a strange disease in Lima in 1638. Her Jesuit confessor, the story goes, gave a medicine to her doctor made from the bark of a common Peruvian tree. It supposedly saved her life and two years later she returned to Spain, carrying with her some of the magic bark. Thus she gave to the world one of the supreme medicines of all times. A century later the Swedish botanist Linnaeus tried to pay a compliment to the long-dead beauty but misspelled her name—calling her tree "cinchona". Out of it came quinine.

The Andean forests remained for 200 years the only source of the magic drug—quinine. The cinchona trees grew wild. They were stripped of bark recklessly and became very scarce. By 1850 the price of quinine was $50 an ounce and only the rich could afford to have malaria.

The British tried to transplant the tree in India and failed. Then Dutch botanists obtained some seed, planted it in the East Indies, and developed high-yielding species. Soon this region became the sole source of the world's supply. The price dropped to 18 cents an ounce and the lands over which the long-dead Countess had ruled dropped out of the picture.

Now South American countries, notably Venezuela and Bolivia, are reclaiming the crop with improved varieties of the cinchona tree, equal to the best produced by the Dutch. They are regaining rapidly the dead lady's gift.

COLOMBIA'S ANT TREE

In the sparsely inhabited, tropical portion of eastern Colombia is an ant tree known as the barrasanta. It is a small, slender tree with showy, red flowers which grows 25 to 30 feet in height. Both trunk and branches are hollow and filled with masses of vicious, biting ants. As soon as the tree is disturbed the insects swarm upon the invader. As a result the tree is generally left alone both by Indians and white settlers. The ants are protected by the branches and in turn protect the host with their fighting prowess.

A curious shrub which grows out of enormous anthills found through the llanos region of western Colombia furnishes quite a different example of insect-plant association. The ants are "leaf cutters." All other plant life avoids their immediate neighborhood. This particular shrub exudes a viscous, milky juice which traps any ants which try to climb toward its leaves. Hence the insects have learned to leave it alone and it enjoys the rich ant hill soil without competition from any other plants.

THE STRANGE BEHAVIOR OF PLANTS

The behavior characteristics of some American plants are strange indeed.

The compass plant, a bristly perennial of the aster family which grows in abundance over the prairies, is a living compass. It turns the edges of its leaves in a general north-south direction. Another American plant, the wild lettuce, does the same thing. The result is that when the intensity of sunlight is weakest in the morning and evening the flat surfaces of the leaves are in a position to receive the maximum available amount of light. At noon, when there is more light than the plant needs, only the edges of the leaves are turned towards the sun.

Then there is the English ivy which arranges its leaves in a mosaic pattern so that about the greatest possible area is exposed to the light. Other plants show equally precise adaptations to their light requirements.

It is all associated with the process of photosynthesis—i.e., the manufacture by the plant of carbohydrates out of carbon dioxide and water in the presence of light. The strength of light needed for this process varies somewhat with the particular plant and its conditions. The phenomenon is one of the most vital in creation, the transformation of the sun's energy into the fuel of animal life. Without it life would be impossible.

Some plants work under high light intensities, such as those which must adapt themselves on the desert areas of the southwestern United States. Others thrive best in the subdued light of a dense forest. One curious little moss grows in caves where there is almost no light at all. It is equipped with a plate of cells forming a battery of lenses capable of focusing the scattered light on the bodies especially concerned in carbohydrate formation. These are the chloroplasts which contain the mysterious substance, chlorophyll, which acts as a catalyst for action of sunlight on carbon dioxide and water. The shape and arrangement of cells containing the chloroplasts are such that the amount of chlorophyll exposed to the sunlight can be varied.

A specially devised apparatus has been constructed in the Smithsonian laboratory for quantitative studies of the way plants absorb carbon dioxide under different lighting conditions. Not only is the process greatly effected by the intensity of the light, the experiments show, but the wave length also is of paramount importance. The experimental plants are grown with their roots in a nutrient solution and their tops extending into a double-walled glass tube. They are furnished light from surrounding lamps, so that the intensity and wave lengths of the light can be varied as desired. Through the tube, air containing different amounts of carbon dioxide can be passed. Thus every element of the process is under rigid control of the experimenters.

The experiment already has shown that the correct combination of wave lengths is of the utmost importance in making up synthetic light. Thus, regardless of the intensity, the ordinary electric light when used alone has been demonstrated to be a poor light source. Its maximum energy occurs in the infrared region, below the limit of visibility, while that of sunlight falls in the green-blue region. If tomato plants are grown under high powered Mazda lamps in the Smithsonian's special growth chambers, especially when the humidity is high, their leaves turn pale and almost white. Chlorophyll disappears under these conditions.

VENEZUELA'S NOCTURNAL ORCHID

A flower that opens only by moonlight is one of Venezuela's plant curiosities. It is an ivory-white, velvety orchid which depends entirely on nocturnal butterflies to sip its nectar while pollenization takes place.

The plant is one of 800 species of Venezuelan orchids. Among these is probably the prettiest and rarest of the orchid family, the mother-of-pearl flower, which can sometimes be found in the deep jungles of the Gran Sabana area at altitudes of more than 3,000 feet.

Still another high mountain variety has square petals with fringed edges. Another, found in the jungles of the Upper Orinoco, has blossoms measuring up to 16 inches in diameter. A unique Venezuelan orchid grows only in water.

Throughout the world there are more than 20,000 species of orchids, the great majority of which are found only in mountainous regions of the tropics. A few, however, grow as far north as the Arctic Circle.

THE PLANT THAT STRIKES MEN DUMB

A plant cultivated in the gardens of the Venezuelan National University at Caracas might well be a boon to pestered husbands and harassed mothers.

It is described under the popular Spanish name of "planta del mudo." It looks like sugarcane. According to the probably exaggerated claims, anybody who chews the stem is stricken dumb for at least 48 hours, presumably due to some paralyzing effect on some part of the vocal apparatus. It is not known whether anybody has tried to extract the marvelous talk-stopping principle.

American botanists are unable to identify the plant. They explain, however, that the northern portion of South America long has been known as the world's greatest storehouse of plants with strange physiological effects. There is one, for example, alleged to grow hair on bald heads, another which makes everything look red.

COMBAT OF MOTH AND SHREW

A strange fight between a grey shrew, smallest of North American mammals, and a black "witch moth" has been described by Laurence M. Huey of the San Diego Society of Natural History.

The moth, with a wing spread of about four inches and a body size almost equal to that of the shrew, was placed in a cage with the mammal. The shrew proved too much for the insect after the odds had been equalized by clipping a great part of the latter's wings.

"Even with this severe handicap", reports Mr. Huey, "the moth still was very strong and, as its body was so large, the shrew attacked it by grasping one of its wing stubs, tugging with main strength, and hanging on like a bulldog. Once, in a burst of spirited action, the shrew was pitched half way across the cage. This only caused a more determined attack and the moth finally was killed and eaten.

"Another moth, with a body about three-quarters of an inch long, was placed in the cage. It had lost many of the scales from its wings and was partially disabled. It could fly feebly, however, from one side of the cage to the other. The shrew, apparently by its sense of hearing, kept following the course of the moth until its flight carried it about two inches above the little mammal. Then, with an almost invisible quickness, the animal sprang and seized the moth in the air, much as a basketball player leaps to catch a ball high over his head. A few crunches with the sharp-toothed jaws dispatched the moth."

THE FEROCIOUS SNAKE WEASEL

From South Africa comes a report from Dr. Raymond B. Cowles of a fight between a deadly reptile and a little known mammal, the inyengelizi, or snake weasel.

The habitat of the snake weasel, unknown in any zoo, is the Umzumbe Valley in Natal Province, where it is one of the rarest of carnivores. Natives either refuse to bring in inyengelizis or demand exorbitant prices for their skins. All parts of the body are used in the native pharmacopoeia and elders wear a narrow strip of the fur to ward off evil and bring good luck.

Little is known concerning the habits of the animal except that it apparently frequents burrows of subterranean animals in gardens, sometimes is ploughed up, and will attack and kill large snakes.

A reliable Zulu described to Dr. Cowles a fight between one of them and a deadly mamba about seven feet long. He said he had been watching the snake, basking in the sun in a coiled position. After a few moments a movement in the bushes caught his attention and he saw an inyengelizi cautiously stealing towards the snake. When within a foot or two the animal suddenly leaped upon the reptile and fastened its teeth just behind the head where it clung during the ensuing wild struggle. After a few minutes it succeeded in killing the snake, whereupon it relinquished its hold, performed its toilet, and left without eating any of its prey.

THE RABBIT THAT SWIMS

Life history and habits of a swimming rabbit are the subject of a report to the American Society of Mammologists. The animal is the little known marsh rabbit of the South Carolina coast. It spends most of its life on the tidal marshes and hence, alone of the rabbit family, has become a partially aquarian animal. Almost strictly nocturnal in its habits, its ways of life hitherto have eluded naturalists.

By far the best known trait of the species is its liking for water. Individuals sometimes are encountered in day time far out in one of the coastal rivers. In summer when the water is warm they take to it readily. They seldom are observed, however, swimming in cold water.

In fall and winter the little animal leads a precarious existence. It is the favorite food of the great marsh hawks, continuously circling over the swamps. When Spring comes the birds leave for the North, the sedges grow tall so as to conceal completely the timid little animals, and they are left in peace until the frosts of Autumn.

Generally the marsh rabbit is a home-loving creature but floods in the fresh water area of its habitat sometimes force a migration. It is a natural swimmer. On land it walks with a swimming motion. Other rabbits are practically helpless in the water and try to swim with the hopping motions they use on land. The rare special type appears to be holding its own in spite of its many enemies.

GORILLA WARRIORS OF THE BELGIAN CONGO

A study of mountain gorillas in a part of the world which they have all to themselves has been reported by Captain C. S. R. Pitman, British zoologist.

The only humans who ever penetrate the dense forests on the Uganda border of the Belgian Congo, where these animals are found, are pigmies, with whom the great apes live on the best of terms. Captain Pitman is one of the few white men ever to have entered the area.

The mountain gorilla is probably the highest of all the gorillas, next to man. One of the two or three ever in captivity was an infant kept at the National Zoological Park in Washington, D. C. Its brain was the largest ever found in an infra-human creature; it almost matched the smallest normal human brains.

Capt. Pitman found the gorilla quite a likeable and peaceful animal. He says:

"Around the male gorilla, on account of its enormous size and strength, coupled in recent years with frequent lapses from grace provoked by unnecessary and undue interference, there has been woven and unfortunately published a fantasy of inaccuracy and exaggeration—so much so that the very homely old male is visualized as an object of dread. The male gorilla, as the family head, is most solicitous for the welfare of his wives and children—a very human trait. On the threat of danger, he accepts full responsibility for the well-being of his charges.

"If the danger is real the females and young are sent off, while the father waits to take on all comers until satisfied that the remainder of the band are out of harm's way. Sometimes, when the danger is sudden and overwhelming, the youngsters are sent up trees to hide until the trouble is over. It is strangely reminiscent of the records of some of the early African explorers relative to tribal customs. When the womenfolk were to be seen busily engaged in their usual vocations in the precincts of a village all was well and no hostility contemplated on the part of the local inhabitants.

"But an absence of women and children was interpreted as unfavorable, signifying that they had been removed to a safe place to enable the warriors to fight unhampered. And so it is with the old male gorilla, for as soon as he bids his family seek safety he is out for mischief, although without direct provocation he is unlikely to attack. There are black sheep in every fold and solitary examples both male and female, which probably have been outlaws for a very good reason, have been known to be abnormally aggressive."

THE BIGGEST "RAT" IN THE WORLD

Close relative of the porcupine, but without quills, is the aquatic coypu, or nutria, of South America. It has become quite valuable in recent years because of its soft fur. Weighing about 20 pounds, it often is referred to as the "biggest rat in the world". It shares with the porcupine large, orange-colored incisor teeth which give it a frightful appearance. Like its barbed northern cousin it is a strict vegetarian, living exclusively on water weeds in its native habitat. Before the last war coypu farms were being established through much of Europe. However some apprehension was felt that it might cause considerable damage to crops if it escaped from its enclosures.

THE SUICIDE MARCHES OF LEMMINGS

Mass death marches of lemmings long have intrigued biologists and psychologists.

The Lapland lemming is a short-tailed animal, related to the meadow mouse, that looks like a miniature rabbit. Through the sub-Arctic winter it lives completely buried under snow through which it burrows in search of mosses and lichens.

It is extremely prolific; females produce two litters of from four to six offspring every year. The numbers soon become far too great to subsist on the sparse supply available in the Scandinavian mountains.

Then, irregularly in periods of from five to ten years, occurs one of the weirdest phenomena of animal life. Acting apparently on a common, subconscious, simultaneous impulse, the entire lemming population starts a mass migration out of the mountains to the lowlands. The animals proceed in a straight line, a few feet apart, each usually tracing a shallow furrow in the soil. They are a devouring scourge, stripping the earth of all vegetation in their path. Their progress seems irresistible. No obstacle stops them. If they come across a man they glide between his legs. If they meet with a haystack they gnaw through it. If a rock stands in their way they go around it in a semi-circle and then resume the straight line of their march. When they come to a lake, river or arm of the sea they swim directly across, vast numbers being drowned on the way. If they encounter a boat they climb over it, so as not to be diverted from a straight line. Curiously, they seem to avoid human habitations. They resist fiercely all efforts to stop them. They will bite a stick or hand, crying and barking like little dogs. Multitudes are destroyed every mile of the way. When the migrating horde reaches the sea it moves straight on—to inevitable destruction.

A few linger behind and eventually make their way back to the mountain habitat. Numbers are so reduced that they are seldom observed. Then a new generation starts and builds up for the next migration.

THE FEROCITY OF THE TIGER

Symbol of ferocity in the animal world is the tiger. When troops of the American 101st Division entered the German city of Halle in 1945 it probably was considered effective psychological warfare tactics on the part of the Nazis to open the zoo cages and let loose the tigers. So far as known, however, the animals did not attack any Americans.

Whether the reputation of the tiger is entirely justified is debatable. "The tiger", says Dr. William M. Mann, long-time director of the National Zoological Park in Washington, "is one of the finest animals that lives. In the cage he is the most snobbish of all aristocrats, his contempt for those who jostle in front of his bars being nothing less than magnificent. He is dignity itself. He condescends to no boyish antics to attract attention as does the chimpanzee, to no begging for sweets as do the bear and elephant, to no pacific, philosophic acceptance of fate such as that of the hippopotamus. You cannot win his favor by a stick of candy. He is above rage or gratitude."

Sometimes adult tigers are captured in traps and sold to circuses. One American circus some years ago had a cage of ten. Their keeper made them perform as another man might spaniels. In the arena they appeared to be a ferocious group. In the menagerie tent, confined in small cages like so many kittens, the keeper could put his hand in their months and rub their teeth. Once he complained bitterly about the tranquility of his charges. "I cannot make a show with ten tame tigers," he argued. "I must have five mean ones to add to the act."

The tiger had a prominent part in the menageries of Indian and Chinese monarchs before the Christian era. It first appeared in Europe about the time of the eastern conquests of Alexander. Well known to the Romans, the animal was one of the most dreaded of all the beasts that appeared in the arena.

Despite its supposed ferocity, no great harm has been done in the few cases in which tigers have escaped from zoos. Often they have returned of their own accord.

THE FEARSOME PORCUPINE

There are more than 1,000 minute barbs on each of a porcupine's many quills. This is the reason why such a quill is very difficult to withdraw from the flesh. The armament of quills, from a half inch to three inches long and developed from hairs of the underfur, renders the "spiny pig" of northern woodlands almost immune to attack. About its only enemy in nature is the giant weasel, the fisher, which has learned the trick of quickly turning the porcupine on its back.

The quills are very lightly attached to the porcupine's body and become detached almost automatically when the creature is attacked. That they can be "shot", however, is almost certainly a fallacy. A victim must actually be in contact with the animal.

THE PLANT THAT STIMULATES VISIONS

In 1560 a Franciscan monk wrote of Aztecs eating a plant called peyotl "which gives them terrible and ludicrous visions, alleviates hunger and thirst, gives strength and incites to battle." It was used, he reported "to bring about a state of ecstasy in which one had prophetic visions."

This was the first known reference in literature to the mescal cactus, *Lopophora williamsii*, whose remarkable effects on the human mind ever since have aroused wonderment. Many have experimented with eating the so-called "buttons" of this cactus and have reported all sorts of terrible and ludicrous visions. But no two experimenters apparently have the same experience. After nearly 400 years the supposed active principle, mescaline, has been extracted and the same effects produced either by swallowing or injection of as little as a half gram.

First comes a decided nausea which lasts about two hours. This passes and is followed by weird hallucinations. One's own body seems distorted, with some parts exceedingly small and some very large. A common experience is the feeling that only one's head is the self. The rest of the body is away somewhere in space. The time sense is badly distorted. Minutes stretch out into hours and days, days and hours are contracted into minutes. There are strange optical delusions—lights flashing before the eyes and floating patches of color. Seldom, however, are actual hallucinatory objects seen.

The consumer has the impression that he thinks more clearly than at other times but it has been found that this thought is based more on the sounds than meaning of words. There is a tendency, for example, to argue in puns. An invisible barrier seems to separate one from the rest of the world. This condition lasts for two or three hours, and then passes away, leaving no after affects. The condition has been likened to schizophrenia.

Large doses produce catatonic conditions. A person may sit motionless for a long time in an apparently quite uncomfortable position and refuse to move. Dogs and cats given mescaline injections crouch motionless in corners of their cages, only rousing themselves from time to time to attack invisible assailants.

It recently has been found that only one chemical constituent of mescaline, beta-phenylethylamine, is responsible for the delusions. This is quite similar in chemical structure to the body hormone adrenaline. There have been conjectures that adrenaline may be changed into the mescaline constituent by some as yet unknown process of body chemistry and that this change may be the physiological cause of schizophrenia.

About 40 years ago a peyotl church was set up by Indians in New Mexico. It followed essentially the Catholic ritual, but with mescal buttons substituted for bread in communion. The U.S. Bureau of Indian Affairs did not interfere with the rites when its investigations indicated that the mysterious drug was not habit-forming and apparently caused no physical injury.

THE PUZZLING PLATYPUS

Fantastic combination of mammal, bird and reptile is the egg-laying, toothless water animal of New South Wales and Tasmania, Australia, the duck-billed platypus. It is clearly a mammal but, with a single exception, it stands quite alone among these warm-blooded animals. The creatures from which it is a survivor probably have been extinct for fifty million years.

It is an animal about twenty inches long from the tip of its horny beak to the end of its broad, flattened tail. It is covered with soft brown fur. Its four legs are short and five-toed. These toes on the front foot are joined by webs like those of aquatic birds which extend beyond the long, sharp, curved toe-nails. On the hind legs of the male are inch-long, sharp spurs through which run minute canals connected with a large gland at the back of the thigh—very much like the poison fangs of a serpent. Yet, so far as can be determined, the gland secretes no poison and the spurs apparently are seldom used in self defense.

The female lays two eggs at a time, each about three-fourths of an inch long and a half inch wide, with strong, flexible white shells. These eggs are not incubated but hatch buried shallowly in sand and straw. The platypus lives on the banks of ponds and quiet streams where it digs burrows as much as 20 feet long with two entrances, one below and the other above the water level. The rear, or land, end of a burrow is enlarged into a small chamber in which the young are reared.

The creatures pass most of the daylight hours asleep in these burrows, curled in rather tight balls. The entrances are concealed in grass and reeds so that the occupants of the burrows are seldom seen. At night the platypus takes to the water. It swims and dives easily and its major food consists of worms and other aquatic animals found in the mud or gravel at the bottom. It has cheek pouches like a squirrel. When it comes up from a dive these pouches are stuffed with the food it has gathered, which is extracted and eaten at leisure.

Adult animals are toothless but in each jaw there is a horny ridge. The young, however, have rootless teeth—a possible clue to their very remote ancestry. Like a bird the platypus has a very small head. There is no division of its brain into two hemispheres, as in all other mammals and most birds. This is a characteristic of the reptile brain.

The creatures can climb with apparent ease. Small groups sometimes are seen sunning themselves on broad tree trunks overhanging the water. They are extremely timid but, when captured, soon become quite tame. In captivity, however, they seldom live long.

The only other member of this animal group is the echidna, or spiny ant eater, of the same part of the world. It is, however, an inhabitant of rocky districts where it digs shallow burrows in sand or hides in rock crevices. The back is covered with sharp, backward-directed spines which give it the appearance of a small porcupine. It has a long, tubular snout from which projects the long, slender tongue covered with some sticky substance. With this it laps up ants and other insects.

Like the platypus, it has short, strong legs with large claws with which it burrows with considerable speed. Burrowing, where possible, is its usual method of flight. Its other defense is to roll itself in a ball, when its sharp spines give it considerable protection. "The only way of carrying the creature", says George Bennett (*Gatherings of a Naturalist in Australasia*) "is by one of its hind legs. Its powerful resistance and the sharpness of the spines will soon oblige the captor, attempting to seize it by any other part of the body, to be relinquish his hold."

Milton Keynes UK
Ingram Content Group UK Ltd.
UKHW030905151124
451262UK00006B/985